Advanced Data Acquisition and Intelligent Data Processing

RIVER PUBLISHERS SERIES IN INFORMATION SCIENCE AND TECHNOLOGY

Consulting Series Editor

KWANG-CHENG CHEN
National Taiwan University
Taiwan

Information science and technology enables 21st century into an Internet and multimedia era. Multimedia means the theory and application of filtering, coding, estimating, analyzing, detecting and recognizing, synthesizing, classifying, recording, and reproducing signals by digital and/or analog devices or techniques, while the scope of "signal" includes audio, video, speech, image, musical, multimedia, data/content, geophysical, sonar/radar, bio/medical, sensation, etc. Networking suggests transportation of such multimedia contents among nodes in communication and/or computer networks, to facilitate the ultimate Internet. Theory, technologies, protocols and standards, applications/ services, practice and implementation of wired/wireless networking are all within the scope of this series. We further extend the scope for 21st century life through the knowledge in robotics, machine learning, cognitive science, pattern recognition, quantum/biological/molecular computation and information processing, and applications to health and society advance.

- Communication/Computer Networking Technologies and Applications
- Queuing Theory, Optimization, Operation Research, Statistical Theory and Applications
- Multimedia/Speech/Video Processing, Theory and Applications of Signal Processing
- Computation and Information Processing, Machine Intelligence, Cognitive Science, and Decision

For a list of other books in this series, please visit www.riverpublishers.com

Advanced Data Acquisition and Intelligent Data Processing

Editors

Vladimír Haasz
Czech Technical University
Prague, Czech Republic

and

Kurosh Madani
University Paris-Est Créteil (UPEC)
Paris, France

River Publishers

Aalborg

ISBN 978-87-93102-73-6 (hardback)
ISBN 978-87-93102-74-3 (ebook)

Published, sold and distributed by:
River Publishers
P.O. Box 1657
Algade 42
9000 Aalborg
Denmark

Tel.: +45369953197
www.riverpublishers.com

Table of Contents

v

1

Introduction

Vladimir Haasz[1] and Kurosh Madani[2]

[1]*Czech Technical University in Prague, Faculty of Electrical Engineering*
[2]*Images, Signals & Intelligent Systems Lab, IUT Senart-FB, University Paris-Est Créteil*

DAQ and data processing is a basic part of all automated production systems, diagnostic systems, watching over quality of production, energy distribution, transport control and in various other areas. Demands on the speed, accuracy and reliability increase in general. It is possible to fulfill these demands not only using superior (nevertheless more expensive) hardware, but also applying advanced data acquisition and intelligent data processing. It deals e.g. with optimal data fusion of a number of sensors, new stochastic methods for accuracy increasing, new algorithms for acceleration of data processing, etc. These are the grounds for publishing this book.

In the first part of the book (chapters 2-5) advanced methods of data acquisition in non-routine applications are presented. It concerns e.g. methods of increasing the vertical and horizontal resolutions of acquired waveforms. The applicability of these methods depends on the nature of the waveform of interest. The best enhancements seem to be achievable for repeatable waveforms; the available measurement time budget can constrain the enhancement obtainable in practice.

The following chapter deals with disaggregation algorithms in conjunction with smart meters providing consumers with detailed statistics of electrical energy consumption. The detailed feedback could lead to a reduction in consumption, including increased consumer awareness of a consumption structure.

The intelligent data fusion enables to achieve good results also with low-cost sensors. An "electronic nose" based on a low-cost array of gas-sensors is presented as an example. Preliminary results show that an intelligent odor-discriminating system based on a gas sensor array can contribute to identification and classification of truffles.

V. Haasz and K. Madani (Eds.), Advanced Data Acquisition and Intelligent Data Processing, 1–3.

Describing of an ultrasonic DAQ system for the ultrasonic transducer properties investigation closes this first part. Discussion on signal and transducer types is presented. Spread spectrum (SS) signals and conventional pulse, step and burst signals were used in comparison. Measured transducer impedance interaction with excitation generator was evaluated for attainable power delivery efficiency and bandwidth.

The second part attends to measured data fusion using up-to-date advanced data processing. The problem of optimal data fusion in the network of stochastic unknown-input observer ensuring the state and input signal estimation for dynamic systems under the presence of stochastic disturbance and random measurement noise is presented in Chap. 6. Optimal stochastic unknown-input observer design method is proposed and the equations for observer optimal parameters are founded as well as the conditions of steady-state observer existence are established. For the problem of state and input estimates fusion in decentralized stochastic observers the optimal estimates aggregation is obtained.

The following chapter deals with detection of an odor in the human living space. It presents an array sensing system of odors and adopts a layered neural network for classification. All measurement data obtained from fourteen metal oxide semiconductor gas (MOG) sensors are applied. In order to classify odors, the data from all fourteen sensors are used even if some of them are not sensitive so much. Three methods of using the data by insensitive sensors to find the features of odors Have been proposed. Then, applying those features to a layered neural network, the classification results are compared.

The last chapter of this part concerns multisensors signal processing. An improved algorithm of intelligent data processing by integrating the modified identification method of individual characteristic curve along with the adaptive neuro-fuzzy inference system (ANFIS) is presented. The results of data prediction with training the ANFIS system and the different number of training epochs are described. The proposed approach provides the high accuracy of data prediction as well as low neural network training error.

The last three chapters of the book are focused on nonlinear systems and multidimensional image processing. Nonlinear sensors and digital solutions are used in many embedded system designs. As the input/output characteristic of most sensors is nonlinear in nature, obtaining data from a nonlinear sensor by using an optimized device has always been a design challenge. A new Adaptive Neuro-Fuzzy Inference System (ANFIS) digital architecture based Field Programmable Gate Array (FPGA) was proposed to linearize the sensor's characteristic. The performance of the developed architecture was examined by comparison to ANFIS software model and to two other designed architectures based on FPGA using classical techniques.

The accuracy of the interpolation method of nonlinear dynamical systems identification based on the Volterra model in the frequency domain is described in the following chapter 10. To extract the n–th partial component in the response of the system to the test signal the n–th partial derivative of the response using the test signal amplitude is found and its value is taken at zero. The polyharmonic signals have been used as the test ones. The algorithmic and software toolkit in Matlab was developed for identification processes. This toolkit is used for constructing the informational model of the test system.

The book is ended by the chapter which relates to multidimensional image processing. By involving the usage of a Cellular Automata (CA) based approach the presented work offers an original policy for overcoming problems relating hyperspectral images' segmentation. The application of CA over the hyperspectral cube leads to homogeneous regions easing the segmentation task. This way of proceeding avoids any form of projection and thus preserves the spectral character of the information in the segmentation process. It is also shown that the evolutionary process ruling the CA can be carried out over RGB images overcoming the need of appropriately labeled ground truth images (required in other usually used techniques). The effectiveness of the proposed approach has been verified as well over synthetic hyperspectral images as on segmentation of real-world hyperspectral images leading to competitive results.

2

Waveform acquisition with resolutions exceeding those of the ADCs employed

He Yin, Mohammad Mani, Omar S. Sonbul and Alexander N. Kalashnikov

Department of Electrical and Electronic Engineering, The University of Nottingham, University Park, Nottingham, NG7 2RD, UK

Abstract

This chapter discusses various software/firmware and hardware methods and architectures to improve the fidelity of the acquired waveforms beyond the vertical and horizontal resolutions that are possible with the ADC employed. The applicability of these approaches, and the limits on the enhancements that are achievable, depend upon the nature of the acquired waveform, and they are presented separately for one-shot, repeatable and repetitive waveforms. The possibilities of combining applicable methods in order to simultaneously increase both resolutions are also discussed. The consideration is illustrated by the simulation results and the acquired experimental waveforms relevant to the ultrasonic non-destructive evaluation.

Keywords: waveform acquisition, vertical resolution, horizontal resolution, analog-to-digital converter (ADC), time-interleaved ADC, averaging, boxcar averaging, synchronous digital averaging, on-the-fly averaging architecture, oversampling, equivalent time sampling, random interleaved sampling, accurate interleaved sampling (AIS), one shot waveform, repetitive waveform, repeatable waveform, dither, two clock AIS architecture, time delayed AIS architecture

V. Haasz and K. Madani (Eds.), Advanced Data Acquisition and Intelligent Data Processing, 5–30.

2.1 Introduction

Waveform acquisition involves digitizing equidistant samples of a signal of interest which high spectral boundary is not negligible compared to the sampling frequency. The collection of the consecutive samples acquired during a particular time interval (time window) is referred to as an acquisition frame. The frame may be displayed for visual inspection, analysis, and measurements, which is common for digital storage oscilloscopes (DSOs); stored or used for subsequent digital processing, feature extraction, and/or parameter estimation in various electronic devices or purpose-built digitizers.

Waveform acquisition can be seen as substituting for the continuous graph of input variable versus time with a collection of associated dots located in the nodes of a graph paper of some scale. Better representation of the continuous waveform of interest can be achieved by increasing the horizontal and/or vertical resolution of the graph paper used. However, any realizable analog-to-digital converter (ADC) imposes particular limitations on both resolutions due to its limited number of bits (more precisely, effective number of bits—ENOB) which determines the vertical resolution and maximum sampling frequency which determines the horizontal resolution commonly referred to as the time base in DSO literature. (These resolutions defined by the ADC itself should not be confused with the horizontal and vertical screen resolutions of DSOs featuring displays.)

The data converter market represents approximately 17% of the world's analogue semiconductor market, which was estimated to be around $20B in 2013 [1]. Additionally, ADCs are built into most microcontrollers, many systems-on-chip and even some field programmable gate arrays (FPGAs) [2]. The costs of the top-of-the-range DSOs can reach six figures (e.g. [3,4]). Every established electronic components distributor offers several hundred types of ADCs and quite a few DSOs that can meet the needs for diverse performance and power and/or cost requirements. Nevertheless, there are several instances when improvement of the ADC/DSO resolutions is highly desirable, for example:

- Most modern low-cost microcontrollers are equipped with built-in ADCs with a resolution ranging from 8 to12 bits and maximum sampling frequencies ranging from 100 kHz to 3 MHz, or more. Whilst these sampling frequencies are more than sufficient for the acquisition of various slowly changing waveforms, the vertical resolutions might not be high enough for achieving the required temperature resolution when converting the output voltage of many analogue sensors, such as temperature [5], force, pressure or humidity [6]. Using an additional higher resolution ADC (or a premium microcontroller with a built-in higher resolution ADC) would increase the cost of the device; however, that could be avoided if the resolution of the built-in ADC was enhanced;

- There are several applications where any increase in the ADC resolutions would be gratefully utilised, most notably DSOs, software-defined radios (SDRs) and high-speed communication links that use some modulation. For example, contemporary DSOs exhibit much higher resolutions compared to the very top-of-the-range ADCs that are achieved by using many lower specs ADCs that are simultaneously integrated into a proprietary integrated circuit as discussed in Section 2.3.2;

- Achieving flexible resolutions by modifying the programmable parts of an existing DSO or relevant software plugins without the need for an expensive hardware re-design or upgrade; and

- Overcoming the limitations imposed by a particular ADC architecture while retaining the benefits (e.g. flash ADCs provide the best horizontal resolutions, but achieving high vertical resolutions is impractical; the opposite applies to sigma-delta ADCs [7, 8]).

This chapter discusses various software/firmware and hardware methods and architectures to improve the fidelity of the acquired waveforms beyond the capabilities of the ADC that is employed. It is organised as follows. Section 2.2 discusses how the ADC resolutions affect the fidelity of the acquired waveforms regardless of their nature. Section 2.3 outlines the methods that are available for enhancing the resolutions for one shot waveforms. The enhancements methods applicable to repetitive waveforms are considered in Section 2.4. Section 2.5 describes the methods for resolution enhancement related to repeatable waveforms. Section 2.6 presents the summary and conclusions.

2.2 What resolutions are sufficient for the task at hand?

2.2.1. Horizontal resolution

Waveform acquisition is founded on the Nyquist sampling theorem (some people prefer to reference Shannon, Whittaker, and Kotelnikov [9]). The theorem states that a band limited signal could potentially be accurately reconstructed if it is sampled with infinite vertical resolution at sampling frequency f_s exceeding the highest frequency of the signal spectrum F_h at least twice. This selection of the sampling frequency will completely prevent spectrum aliasing that distorts the signal otherwise. Strictly, band limited signals started in the Stone Age and must be digitized from this point until the collapse of the Sun. As ADCs came into existence less than 100 years ago and have a limited lifespan, this type of signal cannot be digitized perfectly backwards and forwards nowhere near this scale. Using any time-limited acquisition window in theory extends the spectral content of the signal of interest to infinity, resulting in spectrum distortions through spectrum aliasing. As the full power analog bandwidth of most commercially available ADCs is commonly engineered to be higher than their maximum sampling rate, this aliasing should be reduced by appropriate low pass filtering in the ADC analogue front end (AFE). One of the leading DSOs vendors recommends selecting the AFE bandwidth depending either on the reciprocity of the rise time of pulsate signals of interest [10, p.37] or at up to $5*F_h$ [11, p.13].

DSO manufacturers recommend digitizing a most basic sine wave with a sampling frequency at least 3–5 times its frequency [12], and apply built-in $sin(x)/x$ interpolation to present the digitized waveform with extra clarity [13]. (Although linear interpolation can be used in principle for the acquired waveform reconstruction, it would require 4 times higher sampling frequency to achieve compatible reconstruction results [10, p.38; 11, p.15].) It is not uncommon to see a recommendation of selecting a sampling rate of 10x F_h for embedded designs (e.g., [14]).

Despite sine waves are widely used for testing purposes and quantification of the ADC performance, most electronic devices use digital signals that can be considered as rectangular waveforms. It has been shown that high accuracy measurement of the rise time of a periodic rectangular waveform required a DSO analogue bandwidth that exceeded the fundamental frequency of the waveform by a factor of 20 [15, Fig. 3–6]. Following another recommendation from the same DSO vendor (to keep the DSO's sampling frequency four times higher than its analogue bandwidth [12]), the required sampling frequency eventually exceeded the fundamental frequency (or F_h if applicable) of the signal of interest by a factor of 80. This factor likely presents the upper estimate for the multiplier of the fundamental frequency of the signal because signals with faster transients occupy wider spectra.

Most waveforms of interest can fit between the extreme examples of sine and rectangular waves considered above. There is a viewpoint that increases in the sampling frequency may be unnecessary if the waveform samples acquired at lower frequency are appropriately interpolated (for example, using the *sin(x)/x* function, as mentioned above). We simulated acquisition of a pulse smoothed by a low pass RC-network, as shown in Fig. 2.1, where an ideal switch was in the up position in samples numbered from 20 to 40 only. After processing by a high-order ideal Butterworth filter, the resultant pulse was digitized by several ideal ADCs operating at different sampling frequencies, and interpolated to a virtually continuous waveform using the **resample** function in MATLAB. The interpolation errors calculated relative to the waveform reconstructed from the highest sampling frequency available were found of the magnitude of up to a few percentage points (Fig. 2.2, bottom) that translates to the ENOB after reconstruction of 6 bits only. Although ENOB is referred to the vertical resolution discussed in Section 2.2.2, the obtained value would be unsatisfactory for all but very crude measurements.

Therefore the sampling frequency calculated from the sampling theorem using some reasonable assumption for F_h should be considered as a bare minimum, and must be increased further by a factor ranging from 2 to 40 giving the required sampling frequency f_s of 4..80 F_h depending on the nature of the waveform of interest. It might be overoptimistic to expect the *sin(x)/x* interpolation procedure to compensate for an insufficient sampling rate.

2.2.2 Vertical resolution

Most ADCs produce an output code that is proportional to the input voltage. In the latter, the smallest change that can be resolved depends upon the ADC's reference voltage and the number of bits. As the reference voltage can vary for the majority of the commercially available off-the-shelf ADCs, the ADC resolution is commonly referred to simply as the number of bits in the output code that is fixed by the ADC design.

As an ADC produces a discrete fixed width digital code for the continuous input signal amplitude, there is a difference between the two values that is referred to as a quantization error for a single sample and quantization noise for a waveform. The required ADC resolution (bits) is commonly calculated using the following ubiquitous formula (e.g. [16, section 5]):

$$N_b \geq \frac{SNR(dB) - 1.76}{6.02}, \tag{2.1}$$

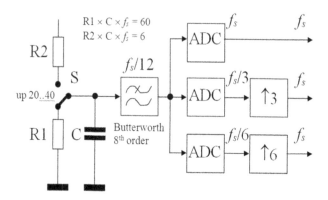

Figure 2.1. Block diagram of the smoothing low pass RC-network and digitizers used for simulating the sin(x)/x reconstruction accuracy versus sampling frequency [49]

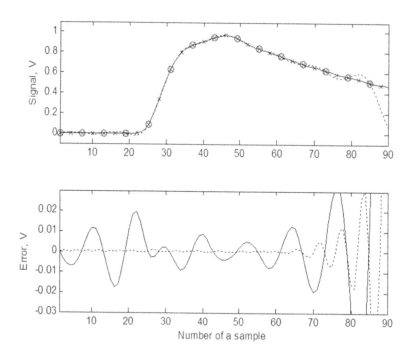

Figure 2.2. Waveforms simulated at different sampling frequencies (top) and reconstruction errors for the fs/6 (solid line) and fs/3 (dashed line) [49]

where the signal-to-noise ratio (SNR) refers to either the input SNR (in this case the calculated N_b ensures that the ADC's quantization noise is lower than the input noise), or to the signal-to-quantization noise ratio (SQNR) alone if the input noise can be considered negligible. (The derivation of a more comprehensive equation that includes some additional factors is presented in [17, eq. 10].)

It is important to remember that (2.1) was derived under some particular assumptions that should be met in order for it to be valid, namely:

- A rounding ADC was considered (truncating ADCs produce more quantization noise because of the offset of half of the ADC least significant bit (LSB); some commercially available ADCs do truncate rather than round their input signals);
- The ADC input is a sine wave (for any other input waveforms it is necessary to consider the ratio between their root mean square (RMS) and the amplitude voltages, known as the loading factor [18], or its reciprocal value, known as the crest factor [19]);
- The amplitude range of the sine wave perfectly matches the ADC input range (in practice, the required gain/attenuation and the offset to the input signal should be provided by the AFE);
- The frequency of the sine wave is not too low to be considered as a nearly direct current for the time window used for the acquisition (otherwise the ADC will produce nearly the same offset for every sample, resulting in a higher RMS overall); and
- The frequency of the sine wave is not too high, as demonstrated below.

Let us consider the importance of the loading factor for several typical waveforms in comparison to the sine wave (Fig. 2.3, top graph), namely an orthogonal frequency division modulated (OFDM) waveform common for telecommunications [20] (Fig. 2.3, middle graph) and an electrocardiogram (ECG) waveform common for biological signal processing [21] (Fig. 2.3, bottom graph). The peak-to-peak amplitude of these waveforms related to their standard deviations is different from that of a sine wave; thus, the realized SQNR should account for the loading factor. By changing the offset voltage of the DSO, one can fit these waveforms into the full operating range of the ADC, and estimate the RMS value of the digitized waveform. The results show substantial deterioration compared to what was expected (Fig. 2.3, the table, best case).

An arbitrary input signal, in addition to offsetting, often requires extra amplification or attenuation to fit the operating range of the ADC.

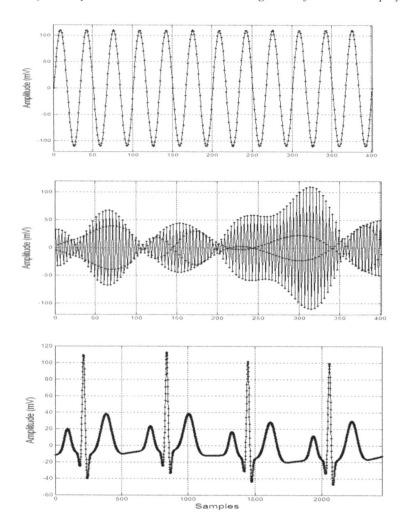

SQNRs CALCULATED FOR THE WAVEFORMS ABOVE

Waveform	SQNR, dB Best case	SQNR, dB Worst case
Sine wave	49.91	41.95
OFDM	44.18	36.22
ECG	42.23	34.27

Figure 2.3. Waveforms used for simulation of the SQNRs and the results obtained
(from top to bottom: sine wave, OFDM, ECG)

Most commonly, the gain can be adjusted according to the following set: 1x, 2x, 5x, 10x, etc if applicable. This means that if the 5x setting is used but the signal even slightly saturates, a lower setting of 2x will have to be used instead. This setting will lead to the reduction of the RMS value of the digitized signal by a factor of up to 5/2=2.5, or 7.96 dB, lowering the worst case QSNR down to 34–42 dBs only, as shown in Fig. 2.3, the table. If the adjustment of the offset was not available, the realized QSNR would even be worse for the ECG case. Therefore it is quite possible to lose around 12 dB QSNR (or 2 bits of ENOB) due to the low loading factor of the input signal and suboptimal gain of the AFE.

To illustrate the importance of the adequate sampling frequency, let us consider sampling a 2-bit binary phase shift coded message with the fundamental frequency of 1 Hz and an amplitude of 1V, presented in Fig. 4a as a solid continuous line, by an ideal sampler with infinite resolution; thus, QSNR=∞ (in MATLAB simulations the actual resolution was 64 bits of the default double precision). In this case, using the sampling frequency of 10 Hz seems appropriate, and the obtained samples are shown in Fig. 2.4a as the stepped solid line that can be used to reconstruct the input signal with the aid of a conventional digital-to-analog converter (DAC) that ideally holds its output voltage intact until the following sample. However, this reconstruction exhibits substantial errors, shown in Fig. 2.4a as the dotted line. The difference between the original waveform and its reconstruction can be significantly reduced if the *sin(x)/x* interpolation is used instead (in Fig. 2.4b, the solid line shows the reconstructed signal, the vertical lines point to the samples taken from the input waveform). The reconstruction error peaks at the bit change instant at about 14% (Fig. 2.4c). The signal-to-reconstruction error ratio (SRER) is 29 dB giving an ENOB of 4.7 bits only according to (2.1). This primarily happens due to the presence of three discontinuities in the waveform that led to profound spectrum aliasing. Nevertheless, even when the input waveform was filtered by an FIR digital filter with the bandwidth of 3 Hz, with a stop band starting from 5 Hz and a side lobe level below -40 dB, the reconstruction error still peaked at about 5% with the overall SRER of 45 dB, which was equivalent to only 7.2 bits according to (2.1). Therefore, the required number of bits calculated from (2.1) should be considered to be a lower estimate for the required number of bits in the general case. If (2.1) is solved for the SQNR, and the latter is calculated for the actual number of bits of the selected ADC, this figure will be achievable in practice if all the conditions mentioned above are valid.

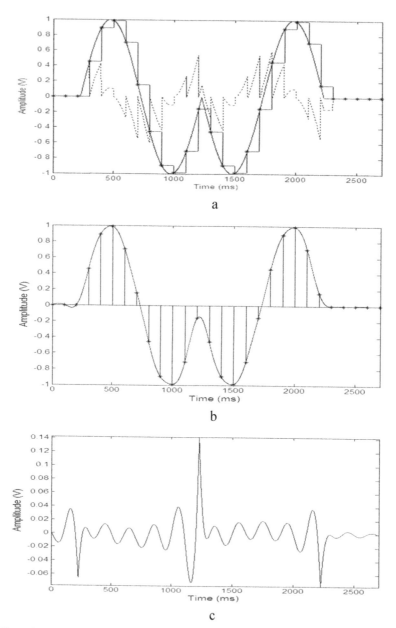

a

b

c

Figure 2.4. Simulation of the reconstruction errors of a 2-bit binary phase shift keying modulated signal sampled at 10x its fundamental frequency by an ADCwith infinite vertical resolution ((a) interpolation by bars provided by common DACs and reconstruction error (dashed line); (b)interpolation by *sin(x)/x*; (c) reconstruction error for *sin(x)/x* (note the scale))

2.2.3   Interrelation between the resolutions

For the successive approximation and for sigma-delta ADC architectures, the output code is produced by iterative processes that generally achieve better vertical resolution given more time. However, increasing the sampling interval decreases the maximum sampling frequency that is available. For these types of architectures, both resolutions need to be considered simultaneously. Some commercial so-called "flexible resolution" DSOs allow trading off one resolution for another (e.g. [22]).

2.3 One shot waveforms

These waveforms relate to one off processes that occur sporadically without repeating or ones that should be acquired for immediate processing or streaming.

2.3.1 Enhancing the vertical resolution

This enhancement is based on the fact that the width required for maintaining the sum of several binary variables without overflow exceeds their width. For example, adding two 8-bit variables requires having a total of 9 bits. If the result of such additions is divided by the number of constituents (the constituents are averaged), in the above example half a bit of resolution can be potentially gained. (This procedure would not work if all the constituents are exactly the same, as the average would equate to the value of the constant samples.) The additional values that are required for enhancement can be obtained by sampling the input signal at a sampling frequency f_{os} that is higher than that required to achieve the desired horizontal resolution by an integer factor k ($f_{os}=k \times F_s$), which is referred to as oversampling. If the oversampled waveforms are processed using moving average whilst keeping f_{os} for the output data, the procedure is known as boxcar averaging [23]. If the processing procedure (that performs either averaging or low pass filtering of the extra samples) eventually reverts the sampling frequency to f_s by decimating the output, vertical resolution enhancement can be achieved. Manufacturers of microcontrollers and DSOs refer to this procedure as oversampling [5, 6], oversampling and decimation [16] or simply resolution enhancement [24]. We will refer to this procedure as oversampling throughout this chapter even though this term can be confused with the procedure applied to reconstruction of analog audio signals using DACs [25]. Additional benefits of oversampling that are useful on their own include easing of the requirements for the AFE anti-aliasing filter and the possibility of filtering out some of the quantization noise [25].

The above consideration can only lead to resolution enhancement if the extra samples are not the same as those mentioned above. In many cases, the input signal features enough noise thereby allowing the oversampling to achieve the desired resolution enhancement without taking any extra measures. However, there are cases when some additional noise should be injected deliberately in order to make the procedure work. (Four methods for implementing injections are mentioned in [5, p.9]). The amount of noise should be high enough to allow for vertical resolution enhancement by oversampling, but low enough so as not to distort the signal of interest very much. The value of at least 1 LSB was suggested in [5, p.13]. We simulated the operation of a rounding digitizer when a random Gaussian noise was added to the input signal. Our study consisted of 1000 samples all kept at zero, and we calculated the RMS value at the output for various oversampling factors (number of averages) (Fig. 2.5). The simulation confirmed that, if the RMS value of the injected noise was below 0.5 LSB, the output RMS value tended to saturate at higher oversampling factors, which meant that the vertical resolution enhancement would not be achieved. For values of 0.5 LSB or higher, the output RMS value followed the theoretical prediction (the output RMS to be proportional to the square root of the number of averages N_A). On this basis, we suggest using the RMS value of 0.5 LSB for the injected noise, if it is required.

To conclude the discussion of oversampling, we need to consider dither (or dithering), which was successfully applied to the low resolution ADCs available for digital audio during the early days of its development. Indeed, dither involved adding noise to the signal of interest before the conversion, but no oversampling was used due to the lack of fast ADCs at that time. This was done in order to smooth out the quantization noise spectrum that otherwise peaked at the harmonics of the fundamental frequency [26]. Despite the increased overall noise level [26], due to the human peculiarities of sound perception, the digitized sine waves sounded much less distorted than these recorded without dither - example can be down-loaded from [27]. Therefore, noise injection for the enhancement of the vertical resolution during oversampling can be called dither, but this should not be confused with the dither that is used for digital audio.

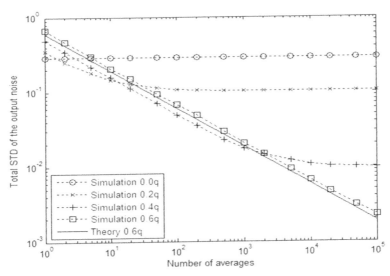

Figure 2.5. Simulated standard deviations of the output noise of a digitizer;
the figure shows different levels of standard deviation of the input additive noise related
to the quantization step (q) of 1 LSB [49]

2.3.2   Enhancing the horizontal resolution

Some ADCs could be overclocked beyond the manufacturer's specification to increase their sampling frequency, but doing this would invalidate their other specified parameters and, nevertheless, it would only marginally extend the maximum sampling rate. A better approach involves getting several identical ADCs to convert the signal of interest in parallel, and then clocking each of them individually to take samples after equidistant delays that cover a nominal minimal sampling interval of an individual ADC. The collected samples are then interleaved to present a final waveform with the sampling frequency increased by the interleaving factor N_I equal to the number of the ADCs used [28]. This hardware architecture is mostly referred to as time-interleaved ADCs (TIADC) [29]. In addition to the increase of the sampling frequency, counter-intuitively this architecture can lead to lower power consumption in comparison to a single high-speed ADC [30]. The most serious drawback of this architecture is the need to compensate for unavoidable mismatches of individual ADC parameters (like offset, gain, time skew [31], AFE and sample-and-hold errors [29]) that are usually software corrected. These mismatches can be reduced to a certain extent if all of the ADCs are implemented on the same semiconductor die. A selection of recently reported research prototypes features TIADCs with N_I ranging from 2 to 128, resolutions ranging from 6-14 bits and sampling at up to 16 Gsps, as shown in [29, Table 2]. Proprietary TIADC designs were successfully employed by various DSO manufacturers; [32] outlines the designs of flagship DSOs for one of the leading DSO vendors spanning from 1987 until 2007, with all the DSOs featuring TIADCs. The DSO built in 2007 contained 80 interleaved ADCs to achieve the overall sampling frequency of 20 Gsps.

It should be noted that ensuring precise timing for individual TI ADCs might be more important for acquisition fidelity than just increasing N_I; some examples of acquired waveforms with poorly time aligned TI ADCs are discussed in [33].

Whilst many DSOs provide multiple input channels by sharing a single high-speed ADC, some DSOs utilise multiple ADCs for different channels. In the latter case, the ADCs can be interleaved when only one channel input waveform is of interest.

2.3.3 Enhancing horizontal and vertical resolutions simultaneously

As TIADCs represent a hardware solution for increasing the horizontal resolution, and oversampling allows for enhancing the vertical resolution in software, they seem to be compatible and can be potentially combined, for example, in order to increase the vertical resolution of a given high-speed ADC.

2.4 Repetitive waveforms

In a strict mathematical sense, a periodic function should perfectly repeat itself after expiration of its time period. Acquired waveforms usually feature some period jitter and additive noise that results in any waveform not being fully compliant to the above definition. For example, a communication bit stream contains level changes happening at the fundamental bit rate but at unpredictable (non-periodic) instants. In electronics, a waveform can be generally considered repetitive if it can be observed on an analog oscilloscope screen, and a bit stream will form an eye pattern (diagram) widely used to establish the appropriate bit rate [34]. The acquired waveform may contain a fraction of a repetition or several repetitions (periods if applicable). Because of the repetitive nature of such a waveform, it is possible to collect several acquisition frames over time and process them to increase the resolutions.

2.4.1 Enhancing the vertical resolution

When several acquisition frames are collected and aligned in the time domain, it is possible to average them sample-by-sample, which is referred to as ensemble averaging [23]. This procedure is commonly used to reduce the additive noise dispersion times the number of averages N_A. In comparison to the resolution enhancement for the one shot waveforms using the oversampling discussed in Section 2.3.1, the samples are collected, not consecutively, but with a substantial delay equal to or exceeding the duration of the acquisition frame which should completely de-correlate any additive noise. Nevertheless, all the considerations of Section 3.1 are applicable here. In particular, increasing N_A leads to $sqrt(N_A)$ reduction of the noise RMS value only if the latter exceeds 0.5 LSB.

The potential of accurate waveform acquisition in the presence of additive noise was demonstrated in the custom digitizer that was capable of acquiring waveforms with amplitudes around 100 nVp-p against a noise background of 335μV RMS using an ADC with the LSB of 488 μV [35]. As noted by one study: "The physical basis of this performance is that the low amplitude input signal is added to electronic noise at the system input and 'rides' on this noise as it passes through the system" [36].

There is one notable distinction that relates to waveform averaging. Perfect aligning of the acquired frames in the time domain is not possible if the ADC and the source of the waveform are not clocked from the same source. This misalignment results in a particular frame jitter noise [34] that needs to be taken into account when assessing the averaged waveform's fidelity.

Most DSOs feature an averaging mode with the maximum number of averages up to 1,000 or more, depending on the DSO model. Sometimes the acquired waveforms with the increased vertical resolution cannot be displayed on the DSO's screen with the full resolution of the former due to the limited number of the pixels of the latter, but it can be stored for future use.

2.4.2 Enhancing the horizontal resolution

If the unpredictable time differences from the trigger event to the sampling edge of the ADC clock are measured with high accuracy, and a number of frames are collected, they can be interleaved in the time domain to reconstruct the waveform. For this procedure to work, the waveform repetitions should be strictly asynchronous to the ADC clock [38]; otherwise, only a limited range of the time differences can be obtained. The reconstruction could consist of analysing the recorded delays and selecting the waveforms that arrived most closely to the equidistant N_I time instants that cover the ADC sampling period [38]. Because the differences are random, the number of the collected acquisition frames is expected to be sufficiently large; achieving N_I=20 requires the collection of 104 frames, on average, but sometimes many more are needed [39]. Additionally, *sin(x)/x* interpolation over the selected set may be used in order to improve the quality of the acquired waveform [40]. This procedure is referred to as random interleaved sampling (RIS) [38] or random equivalent time sampling (ETS) [40]. If the time differences are not measured accurately, interleaving the acquired frames may lead to numerous glitches due to errors in the time alignment (e.g. [41, Fig. 11b]).

Another approach is to progressively delay the sampling edge of the ADC clock relative to the trigger event for every subsequent repetition of the waveform of interest by a specific and precise amount. A single sample may only be acquired during the entire repetition if the repetition frequency is commensurate with the ADC clock frequency and the generated internally synchronous trigger is used to clock the ADC directly [42, 43]. Alternatively, the precise inter-repetition delays may be applied to the continuing ADC clock pulses in order to acquire many samples over every repetition [44]. In this case, the number of acquired waveforms does not need to exceed N_I, which reduces the required acquisition time in comparison to the RIS. Confusingly, this procedure may also be referred to as ETS [43].

Regardless of whether or not the asynchronous or synchronous trigger is used for ETS, all the referenced studies stress that it must be stable; thus, it must occur at the same time instance for every repetition. Trigger stability was simulated using the first order triggering model [44, Section 5]; the simulation showed that the additive noise introduces jitter to the otherwise stable trigger event and the jitter RMS value increased with the decrease in the gradient of the trigger signal for the same RMS value of the additive noise. Therefore, noisy waveforms with slow transients may not be suitable for the ETS waveform acquisition.

2.4.3 Enhancing horizontal and vertical resolutions simultaneously

In general, combining ensemble averaging with RIS for repetitive waveforms seems rather complicated and it will likely require that a massive number of frames be acquired with the stable trigger.

2.5 Repeatable waveforms

Repeatable waveforms can be started at any time by the excitation signal from the instrument. They are essential, for example for distance measurement, object classification and remote sensing using radars and sonars; and, for measurement of the parameters and fault testing of the electrical and electronic devices, for example time-domain reflectometry (TDR) and ultrasonic non-destructive evaluation (NDE).

2.5.1 Vertical resolution

Operating repeatable waveforms makes it possible to tightly synchronise the excitation pulses with the ADC clock, eliminating most of the frame jitter noise [45]. The remaining frame jitter relates to the long-term drift of the frequency of the master oscillator [45]. If necessary, this drift can be reduced by using an atomic oscillator or oscillators that are temperature compensated, oven controlled or GPS-disciplined instead of a more common crystal oscillator [46]. As discussed in the previous sections related to the vertical resolution enhancement (Sections 2.3.1 and 2.4.1), some additive noise should be present in the waveform at the ADC input to enable this enhancement.

Fully synchronous digital averaging (SDA) can be implemented by updating the running totals for every waveform sample by adding the newly acquired frame samples to the totals in the software. This procedure would increase the overall time required for the fast waveform acquisition because it takes a considerable amount of time to update the totals for the waveforms acquired from many samples. During this update, the acquisition of extra frames should be postponed so as not to overwrite the most recently acquired frame until it is fully processed. A better alternative for SDA implementation is to update the running totals as soon as every new waveform sample becomes available from the ADC by employing dedicated hardware (on-the-fly averaging architecture [45]). This architecture allows for reducing the waveform acquisition time to its theoretical minimum of:

$$t_{acq} = \frac{N_A}{F_R},$$
(2.2)

where F_R is the excitation repetition frequency.

Because SDA allows for the elimination of most of the frame jitter noise, better acquisition of waveform fidelity can be achieved in comparison to the ensemble averaging method applied to repetitive waveforms, as discussed in Section 2.4.1.

2.5.2 Horizontal resolution

As the excitation and ADC clock pulses are derived from the same master clock for repeatable waveforms, and because they no longer need to be derived from the waveform of interest as was the case for RIS, the acquisition trigger becomes as stable as one can obtain from contemporary instrumentation that enables accurate interleaved sampling (AIS). AIS makes it possible to acquire only N_I subsequent waveforms with time delays that are precise to each other without the need for any additional frames to implement RIS. When compared to TI-ADCs, AIS eliminates any mismatches among individual ADCs as only the same ADC is used for the acquisition of all the frames that are interleaved.

The precise delays required to implement AIS can be obtained using either two separate free running oscillators derived from the master clock (two clock AIS architecture [41]) or a delay line formed by a particular quantity of cascaded logic gates (time-delayed AIS [47]).

2.5.3 Enhancing horizontal and vertical resolutions simultaneously

Unlike the case for the repetitive waveform acquisition, acquiring repeatable waveforms does allow for combining both resolution enhancements at the expense of the increased waveform acquisition time:

$$t_{acq} = \frac{N_a \times N_I}{F_R}.$$
(2.3)

Even for excitation repetition frequencies in the low kHz range (the 1 to 10 kHz range is commonly used for ultrasonic NDE), the required waveform acquisition time may increase to a few seconds and even tens-of-seconds. Balancing N_A and N_I values within the available measurement time budget of the process of interest and the frequency of the master oscillator do not vary notably throughout the complete acquisition time.

A combination of on-the-fly averaging and AIS architectures can be implemented using an FPGA, as was done for several revisions of an ultrasonic NDE digitizer in our laboratory [49]. Fig. 2.6 presents the ultrasonic waveforms acquired in the pulse echo mode from a 20 MHz transducer using an 80 MHz clocked ADC that illustrates increases in the acquired waveform fidelity for various combinations of N_A and N_I [50]. The right side of Fig. 2.6 shows that the acquired waveform slightly drifted in the time domain from one acquisition to another due to changing ambient temperature. Such a high achieved horizontal resolution can be exploited to measure temperature ultrasonically [51].

2.5.4 Application examples for ultrasonic NDE

Ultrasonic NDE is used to test the quality of finished products and to monitor the manufacturing processes and image structures of biological objects *in vivo*. These applications differ by the available measurement time budget: assessment of finished articles is very restricted, whilst observing evolving and live processes imposes particular limitations. These limitations are defined by the speed of evolution, as the repeatable signal must not vary much during the complete acquisition time, and by the need to extract the relevant information out of the acquired waveforms, which requires some processing time.

Examples of using acquired ultrasonic waveforms for quality control of hardened steel samples are presented in [52]. This application required substantial values for both N_A and N_I because high ultrasound attenuation in the steel samples lowered the input SNR, and the ultrasound velocity needed to be estimated with rather high resolution [52]. The TDR measurements discussed in [48, Section 4] were characterised by quite low additive noise and they required high time domain resolution, thus, resulting in lower N_A but higher N_I. In the case in which liquid foodstuffs are evaluated for consumption safety, both ultrasound attenuation and velocity were not that high, which allowed for the use of moderate N_A and N_I values [53].

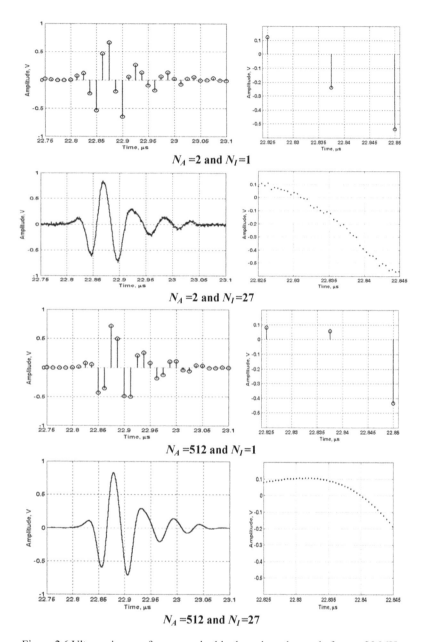

$N_A =2$ and $N_I=1$

$N_A =2$ and $N_I=27$

$N_A =512$ and $N_I=1$

$N_A =512$ and $N_I=27$

Figure 2.6.Ultrasonic waveforms acquired in the pulse echo mode from a 20 MHz trans-ducer using an 80 MHz clocked ADC for various combinations of N_A and N_I [50, figs. 9, 10]

The measurement time budget was not found to be very restrictive when the formation of bio-compatible scaffolds in a stainless steel reactor was monitored because the manufacturing process took a substantial amount of time [54]. However, the monitoring of chemical reactions required faster waveform acquisition and processing [55] that was achieved by utilising a processor embedded into the FPGA fabric [56].

Finally, ultrasonic measurements of the intima-media thickness (IMT) of the common carotid artery *in vivo* were found to be the most restrictive. Despite the low input SNR that resulted from weak ultrasound reflections from tissue interfaces in the human body, it was possible to use only very few averages because providing higher N_I became more important for estimating the thickness with high resolution [57].

2.6 Summary and conclusions

This chapter discussed various software/firmware and hardware methods and architectures to improve the fidelity of the acquired waveforms beyond the capabilities of the ADC that is employed by enhancing their vertical and horizontal resolutions, on their own and simultaneously.

Section 2.2 considered which of the horizontal and vertical resolutions are required to achieve high fidelity waveform acquisition. The literature survey indicated that the waveform sampling frequency might need to be as high at 40 times the Nyquist sampling frequency for accurate measurements of the transient times. It was shown by simulation that *sin(x)/x* interpolation widely used for waveform reconstruction could not appropriately compensate for insufficient sampling frequency in the generic case of communication waveform acquisition that was considered. Other simulations showed that eq. 1, which is commonly used for assessment of the required number of ADC bits, needs to be used with care as it was derived for specific operating conditions that are not valid in a general case.

Sections 2.3–2.5 discussed methods for the enhancement of horizontal resolutions, vertical resolutions and both resolutions, simultaneously, depending on the nature of the waveform of interest; namely one-shot, repetitive and repeatable waveforms, respectively. The findings are summarised in Table 2.1 below. Counter-intuitively, enhancement of the vertical resolution might require injecting extra additive noise at the input of the ADC if that resolution is to occur. It was shown that the best resolution enhancements are achievable for repeatable waveforms at the expense of the increased waveform acquisition time.

Table 2.1

	Vertical resolution	**Horizontal resolution**
General	Every bit of the ADC resolution increases SQNR by 6.02 dB. The exact SQNR value depends upon the parameters of the acquired waveform.	The sampling frequency should exceed the fundamental frequency of the acquired waveform (or its high spectrum boundary if applicable) by the factor of 4 to 80. *Sin(x)/x* interpolation might not compensate for the insufficient sampling frequency.
One-shot waveforms	Can be enhanced by oversampling, provided that enough additive noise is present at the input. If the noise is not present, it should be injected deliberately. We recommend using the noise RMS value slightly above 0.5 LSB.	Can be enhanced by using time-interleaved ADCs. Individual ADC mismatches should be reduced and/or compensated. Precise clocking of individual ADCs is required.
	Combining the enhancements is possible.	
Repetitive waveforms	Can be enhanced by ensemble averaging if some additive noise is present, as stated in the entry above. Additionally reduces the additive noise RMS value.	Can be enhanced by random interleaved sampling (or equivalent time sampling) if the acquired waveform does not change much from one repetition to another during the entire acquisition time. Requires a substantial amount of measurement time and a stable trigger.
	Combining the enhancements seems very complicated.	
Repeatable waveforms	Can be enhanced by using synchronous digital averaging.	Can be enhanced using accurate interleaved sampling.
	Combining the enhancements is straightforward at the expense of the increased acquisition time.	

References

[1] S.Elder, "Does the price of analog matter?", available on http://tinyurl.com/pl5o2lz, accessed Jan 2014.

[2] C.Murphy, "Driving the Xilinx Analog-to-Digital Converter", Application note XAPP795 (Xilinx), accessed online on http://tinyurl.com/nodoglr, accessed Jan 2014.

[3] "Tektronix 33-GHz scopes handle 100 Gsamples/s", available online on http://tinyurl.com/pfjg87q, accessed Jan 2014.

[4] S.Shahramian, "Experiments and demo of an Agilent DSA-X 96204Q 160GS/s 62GHz oscilloscope", available online on http://tinyurl.com/nezn3qt, accessed Jan 2014.

[5] "AVR121: Enhancing ADC resolution by oversampling", Application note (Atmel), available online on http://tinyurl.com/n45srm7, accessed Jan 2014.

[6] J.M.Madapura, "Achieving higher ADC resolution using oversampling", Application note (Microchip), available online on http://tinyurl.com/nwqaw3m, accessed Jan 2014.

[7] The data conversion handbook, W.Kester (ed.), available online on http://tinyurl.com/8swzw, accessed Jan 2014.

[8] "Analog-to-digital converter", Wikipedia, available online on http://tinyurl.com/32zsnl, accessed Jan 2014.

[9] M.Unser, "Sampling – 50 years after Shannon", Proc.IEEE, vol.88, No 4, pp.569-587.

[10] "XYZs of oscilloscopes", primer (Tektronix), available online on www.tek.com after registration, accessed Jan 2014.

[11] "Fundamentals of signal integrity", primer (Tektronix), available online on www.tek.com after registration, accessed Jan 2014.

[12] "Evaluating oscilloscope sample rates vs sampling fidelity", application note (Agilent), available online on http://tinyurl.com/ba8he3c, accessed Jan 2014.

[13] C.Rehorn, "Sin(x)/x interpolation: an important aspect of proper oscilloscope measurements", available online on http://tinyurl.com/bywksa5, accessed Jan 2014.

[14] J.W.Valvano, Embedded Systems: Real-Time Operating Systems for the ARM Cortex M3, 2012, p.95.

[15] "Evaluating oscilloscope bandwidth for your application", application note (Agilent), available online on http://tinyurl.com/b7t3brj, accessed Jan 2014.

[16] "How many bits are enough? The trade-off between high resolution and low power using oversampling modes", application note AN4075 (Freescale Semiconductor), available online on http://tinyurl.com/qebtbpx, accessed Jan 2014.

[17] "Improving ADC resolution by oversampling and averaging", application note AN118 (Silicon Labs), available online on http://tinyurl.com/p6m6xn2, accessed Jan 2014.

[18] R.G.Lyons, Understanding digital signal processing, Ch.12, 13. Prentice-Hall, 2004.

[19] "Crest factor", Wikipedia, available online on http://tinyurl.com/36mzlzg, accessed Jan 2014.

[20] G.Acosta,"OFDM simulations using Matlab", Georgia Inst.Techn., 2000, available on http://tinyurl.com/amr33y9, accessed Jan 2014.

[21] "ECGSYN: a realistic ECG waveform generator", available online on http://tinyurl.com/b4v5bev, accessed Jan 2014.

[22] "Flexible resolution oscilloscopes", Pico technology, available online on http://tinyurl.com/oxvjcn7, accessed Jan 2014.

[23] D.A.Skoog, F.J.Holler and T.A.Nieman, Principles of instrumental analysis, 5th edition, Thomson Learning, 1998, ch.5C-2.

[24] "Resolution enhancement", technical note (Pico technology), available online on http://tinyurl.com/qal9rwn, accessed Jan 2014.

[25] "Oversampling", Wikipedia, available online on http://tinyurl.com/23t7nd7, accessed Jan 2014.

[26] N.Aldrich, Dither explained, available online on http://tinyurl.com/6gbfdy, accessed Jan 2014.

[27] "Dither", Wikipedia, available online on http://tinyurl.com/3aymr4, accessed Jan 2014.

[28] W.C.Black and D.A.Hodges, "Time interleaved converter arrays", *IEEE J. Solid-State Circuits*, vol.15, no.6, pp.1022-1029, Dec 1980.

[29] C.R.Parkey and W.B.Mikhael, "Time interleaved analog to digital converters: tutorial 44", *IEEE Instr. Measur. Magazine*, Dec 2013, pp.42-51.

[30] D.Draxelmayr, "A 6b 600 MHz, 10 m W ADC array in digital 90 nm CMOS", *IEEE Int. Solid-state circuits Conf.*, pp.14.7, Feb 2004.

[31] D.G.Naim, "Time- interleaved analog-to-digital converters", *IEEE Custom Integrated Circuits Conf.*, pp.289-296, 2008.

[32] J.Corcoran and K.Poulton, "Analog-to-digital converters: 20 years of progress in Agilent oscilloscopes", *Agilen Meas. J.*, iss.1, 2007, pp.34-40.

[33] J.Hancock, "Measuring oscilloscope sampling fidfelity", *Agilen Meas. J.*, iss.1, 2007, pp.28-33.

[34] "Eye pattern", Wikipedia, available online on http://tinyurl.com/2q7zrp, accessed Jan 2014.

[35] A.P.Y.Phang, R.E.Challis,V.G. Ivchenko and A.N.Kalashnikov, "A field programmable gate array-based ultrasonic spectrometer", *Meas. Sci. Technol.*, vol.19, id 045802, 13p., available online on http://tinyurl.com/o72ljcv, accessed Jan 2014.

[36] R.E.Challis and V.G.Ivchenko, "Sub-threshold sampling in a correlation-based ultrasonic spectrometer", *Meas. Sci. Technol.*, vol. 22, id 025902, 12 p., available online on http://tinyurl.com/q5yp4ym, accessed Jan 2014.

[37] A.N.Kalashnikov, R.E.Challis, M.U.Unwin and A.K.Holmes, "Effects of frame jitter in data acquisition systems", *IEEE Trans. on Instrum. Measur.*, vol.54, iss.6, pp. 2177 - 2183, Dec 2005, available online on http://tinyurl.com/ntovl9b, accessed Jan 2014.

[38] RIS available online on , accessed Jan 2014.

[39] "Timebase sampling modes", a chapter from a user manual (LeCroy), available online on http://tinyurl.com/ozbg6qf, accessed Jan 2014.

[40] "Real-Time Versus Equivalent-Time Sampling", technical note (Tektronix), available online on http://tinyurl.com/oor4ck5, accessed Jan 2014.

[41] V.Ivchenko, A.N.Kalashnikov, R.E.Challis and B.R.Hayes-Gill, "High-speed digitizing of repetitive waveforms using accurate interleaved sampling", *IEEE Trans. Instrum. Measurem.*, vol.56, iss.4, p.1322-1328, 2007, available online on http://tinyurl.com/lmdzjxn, accessed Jan 2014.

[42] Y. Zheng and K. L. Shepard, "On-chip oscilloscopes for non-invasive time-domain measurement of waveforms in digital integrated circuits", IEEE Trans. Very Large Scale Integr. (VLSI) Syst., vol. 11, no. 3, pp. 336–344, Jun. 2003.

[43] "What is the difference between an equivalent time sampling oscilloscope and a real-time oscilloscope?", application note (Agilent), available online on http://tinyurl.com/6eut79h, accessed Jan 2014.

[44] "DS100 oscilloscope manual", product manual (IBZ Electronics), available online on http://tinyurl.com/nhntdvc, accessed Jan 2014.

[45] A.N.Kalashnikov, "Waveform measurement using synchronous digital averaging: Design principles of accurate instruments", *Measurement*, vol.42, pp.18-27, 2009.

[46] H.Zhou, C.Nicholls, T.Kunz and H.Schwartz, "Frequency accuracy & stability dependencies of crystal oscillators", technical report SCE-08-12, Nov 2008, available online on http://tinyurl.com/p9hg8lk, accessed Jan 2014.

[47] A.Afabeh, He Yin and A.N.Kalashnikov, "Implementation of accurate frame interleaved sampling in a low cost FPGA-based data acquisition system", *2011 IEEE 6th Int. Conf. on Intelligent Data Acquisition and Advanced Computing*, pp.20-25.

[48] He Yin, M.Mani, O.Sonbul and A.N.Kalashnikov, "Measurement time as a limiting factor for the accurate acquisition of repeatable waveforms", *2013 IEEE 7th Int. Conf. on Intelligent Data Acquisition and Advanced Computing*, pp.6-9.

[49] He Yin and A.N.Kalashnikov, "An electronic architecture for intelligent portable pulse-echo ultrasonic instrument", *2nd Int. Conf. Advanced Information Syst. Technol. AIST-2013*, pp.119-120, available online on http://tinyurl.com/ofgbrad, accessed Jan 2014.

[50] A.N.Kalashnikov, V.Ivchenko, R.E.Challis and B.R.Hayes-Gill, "High accuracy data acquisition architectures for ultrasonic imaging", *IEEE Trans. Ultrason., Ferroel. Freq. Contr.*, vol.54, pp.1596-1605, 2007, available online on http://tinyurl.com/nn5kl5j, accessed Jan 2014.

[51] A. Afaneh, S. Alzebda, V. Ivchenko, and A. N. Kalashnikov, "Ultrasonic Measurements of Temperature in Aqueous Solutions: Why and How," *Physics Research International*, vol. 2011, article ID 156396, 10 pages, 2011, available online on http://tinyurl.com/ne7nsjq, accessed Jan 2014.

[52] W.Chen, A.N.Kalashnikov, R.E.Challis and M.G.Somekh, "Experimental ultrasonic assessment of steel induction hardening by measuring two distinct times of flight", *Universal Journal of Materials Science*, vol.1, no.4, pp.201-209, 2013, available online on http://tinyurl.com/ntgewmc, accessed Jan 2014.

[53] HeYin, A.Afaneh and A.N.Kalashnikov, "Discriminating samples of drinkable water by their ultrasound time-of-flight (TOF)", presented at 2013 IEEE Ultrasonics Symposium, publication pending.

[54] M.L.Mather, J.A.Crowe, S.P.Morgan, L.J.White, A.N.Kalashnikov, V.G.Ivchenko, S.M.Howdle and K.M.Shakesheff, "Ultrasonic monitoring of foamed polymeric tissue scaffold fabrication", *J. of Materials Science: Materials in Medicine*, Vol.19, no.9, pp.3071-3080, 2008.

[55] A.N.Kalashnikov, K.L.Shafran, V.G.Ivchenko, R.E.Challis and C.C.Perry, "In situ ultrasonic monitoring of aluminum ion hydrolysis in aqueous solutions: instrumentation, techniques and comparisons to pH-metry", *IEEE Trans. Instrum. Measurem.*, vol.56, no.4, pp.1329-1339, 2007, available online on http://tinyurl.com/nn5kl5j, accessed Jan 2014.

[56] A.Afaneh and A.N.Kalshnikov,"Embedded processing of acquired ultrasonic waveforms for online monitoring of fast chemical reactions in aqueous solutions", In: V.Haasz, ed., Adavanced distributed measuring systems: exhibits of application, River Publishers, pp.67-93, 2012.

[57] M.Mani and A.N.Kalashnikov, "In vivo verification of an intelligent system for accurate measurement of intima-media thicknesses", *2nd Int. Conf. Advanced Information Syst. Technol. AIST-2013*, pp.119-120, available online on http://tinyurl.com/ofgbrad, accessed Jan 2014.

Biographies

He Yin completed his third year of study at Beihang University in China (formerly known as BUAA) in 2008. Later that same year, he began studying at the University of Nottingham as an exchange student. In 2010 he was awarded a MEng degree in Electronic Engineering from the University of Nottingham as well as a BSc degree from Beihang University. He is currently writing his PhD thesis, "Low cost portable instrumentation for high temporal resolution ultrasonic NDE of liquid objects," at Nottingham University.

Mohammad Mani was awarded a MEng degree in Electronic and Computing Engineering at the University of Nottingham, UK in 2008. At present, he is finalizing his PhD thesis, "Ultrasonic instrument for accurate measurements of spatial parameters in blood vessels," at Nottingham University.

Omar S. Sonbul was awarded a BSc degree in Electrical and Computer Engineering in 2003 from Umm Al-Qura University in Makkah, Saudi Arabia. He then completed his MSc degree in Electronic Communications and Computer Engineering at the University of Nottingham, UK. He was later awarded a PhD degree at the same university for his thesis, "Intrusion detection and profile imaging using networked electronic modules for air coupled ultrasonic transducers." He is currently Chairman of the Computer Engineering Department in the College of Computer and Information Systems at Umm Al-Qura University. His research interests include wireless and embedded instrumentation development for air-coupled ultrasonic applications.

Alexander N. Kalashnikov (PhD, DSc, SM IEEE) was awarded an engineering diploma in electroacoustics and ultrasound technology (five-year degree program) in 1984 and a PhD degree in 1991 at the Odessa State Polytechnic University (in the former USSR). He was awarded a DSc degree in 2002 from the same university (Ukraine). Dr. Kalashnikov began working in the UK in 1999 and is currently Associate Professor of Electronic Architectures and Embedded Computing at the University of Nottingham. In 2007 he was elevated to the grade of Senior Member in the Institute of Electrical and Electronic Engineers (IEEE, USA). His research interests include digital signal processing and instrumentation development for applied ultrasonic technology.

3

Different appliance identification methods in non-intrusive appliance load monitoring

Krzysztof Liszewski[1], Robert Łukaszewski[1], Ryszard Kowalik[2], Łukasz Nogal[2] and Wiesław Winiecki[1]

[1]*Faculty of Electronics and Information Technology, Institute of Radioelectronics, Warsaw University of Technology*

[2]*Faculty of Electrical Engineering, Institute of Electrical Power Engineering, Warsaw University of Technology*

Abstract

The aim of appliance identification methods is to get electrical energy consumption at appliance level based on aggregate measurements from a single energy meter. The appliance identification methods in conjunction with smart meters provide consumers with detailed statistics of electrical energy consumption. The detailed feedback could lead to a reduction in energy consumption, including increased consumer awareness. In this paper we describe electrical parameters and appliance groups with different electrical behaviors. We review different steady and transient state identification methods. Then we compare requirements and limitations of discussed methods.

Keywords: electrical appliance identification, smart metering

3.1 The necessity for energy monitoring systems

Growing prices of electrical energy increase the demand for development of systems monitoring and managing electrical energy consumption. European Union noticed the problem and included the need for smart metering systems and smart grids in its directive. Currently, plans for

V. Haasz and K. Madani (Eds.), Advanced Data Acquisition and Intelligent Data Processing, 31–58.

Advanced Metering Infrastructure (AMI) systems designed by energy suppliers do not include detailed analysis of energy consumption by consumers in households. Providing consumers with detailed statistics of electrical energy consumption could increase consumer awareness and lead to a reduction in energy consumption [1]. Thus, it becomes important to develop new methods for monitoring energy consumption more adapted for end-users. The solution of using a measuring energy consumption meter for each device is inefficient. Whereas, the Non-Intrusive Load Monitoring Systems (NIALMS) offer a possibility to monitor all appliances based on aggregated data from a single meter. Main advantages of such systems are the relatively simple installation process, low cost of the installation and small requirements for modification in existing electrical installations. NIALMS provides an estimated electricity consumption of individual appliances. The main aim of the work on various methods of appliance identification is to develop non-intrusive simple methods that will allow to obtain an estimate electricity consumption possibly as close as the actual energy consumption.

The work on non-intrusive monitoring systems was started in 1980s by George Hart [2]. Since the first publication many different appliance identification methods have been proposed. The main purpose of this chapter is to present different identification methods and to classify them into groups based on variety of parameters.

3.2 Energy monitoring systems

Electrical energy monitoring systems can be divided into intrusive and non-intrusive.

The easiest way to monitor electrical energy consumption by a single appliance are Intrusive systems (IALMS-Intrusive Appliance Load Monitoring System). For example, in such systems individual meters are used for each appliance or an unique signal is transmitted indicating current state of the appliance. Therefore, in intrusive systems the use of special appliance identification methods is not required and appliances with the same or similar signatures (including lighting in different rooms) can be identified. On the other hand the intrusive systems are inefficient because of high installation costs as a large number of meters or modules transmitting information about appliance state is necessary. The introduction of a new appliance in such system requires a new meter and system reconfiguration.

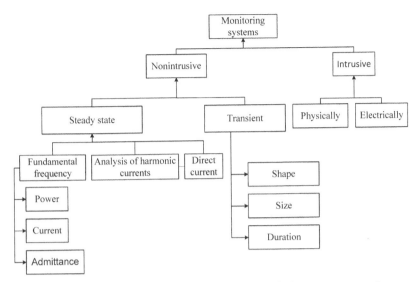

Figure 3.1 Classification of methods for monitoring electrical energy consumption

The second group of systems monitoring energy consumption by separate appliances are NIALMS systems. These systems estimate individual appliance energy consumption using aggregate measurements from a single main meter and electrical appliance signature for each appliance. The appliance signature is a set of parameters describing electrical nature of the appliance obtained from passive observation of these parameters during normal work cycle of the appliance. The signature may contain information about values of electrical parameters during steady or transient states. The information about transient state can specify shape and duration of changes in electrical parameters values. Whereas, in case of steady states, values of active power, reactive power, current or admittance may be used to identify appliances [2].

The main advantage of NIALMS systems is that these systems provide electrical energy consumption of each appliance calculated from measurements from a single central meter. In addition introduction in the home area a new appliance does not require a new measuring meter. The disadvantage of the NIALMS systems is the necessity of using advanced signal processing algorithms in order to identify appliances.

3.3 Electrical energy consumption appliances

Electrical devices, that are under evaluation in case of NIALMS systems, are devices using electrical energy supplied in a form of alternating current 3 voltages through 3 phase solidly earthed network, called very often small voltage distribution network. Depending of particular country, nominal phase voltage (measured between phase conductor as one point and ground conductor as second point) would have 230Vac or 110Vac.

Devices depending of their maximum nominal power are constructed as single phase, 2 or 3 phase. Usually devices with nominal power smaller than 5kW (in case of 230V nominal voltage of network) are constructed as 1 phase, while devices with nominal power above mentioned level is constructed as 3 phase. Figure 3.2 presents the supply circuits of 3 and 1 phase energy consumption devices.

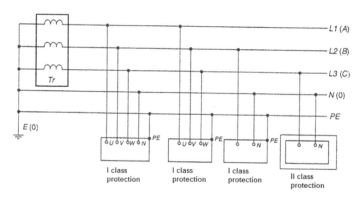

Figure 3.2 Electrical energy consumption devices supply circuits

Above mentioned, devices may be divided into many groups taking into account different criteria.

In case the construction is considering as criterion, it is possible to distinguish simple and complex electrical devices. Complex ones will be made of couple (e.g. two in case of simple heater compose of heating element and motor fan) up to many (e.g. seven or more in case of complicated washing machine compose of heating element, main motor, 2 or 3 motor pumps, switching power supply, microprocessor controller etc) simple electrical devices. It is worth to emphasize that simple machines have different working principles and so influence also different, the 3, 2 or 1 phase power supply and that can be how they can be detected by NIALMS systems.

Considering working principle and power supply influence, electrical devices can be divided into at least following groups:

- incandescent lamp (classic light bulb),
- fluorescent lamp (fluorescent tube),
- compact fluorescent lamp CFL (energy-saving light),
- transformer power supply,
- switching power supply,
- electrical heaters,
- electric cookers,
- induction cookers,
- microwave ovens,
- electric motors (one phase and 3 phase),
- welding machines.

As can be seen above mentioned information, there are many simple electrical devices that can work in household. Each type of device can be characterized by supply voltage and number of phases, maximum power of devices, typical nominal power of devices, inrush current, starting (inrush) time, power factor during normal work, harmonic content of the current with respect to fundamental as can be seen on Table 3.1.

As has been described at the beginning of the chapter 3.3. in case the construction is considering as criterion, each of complex electrical devices can be described as set of basic types presented on Table 3.2.

According to one of distribution companies in Poland typical receivers of electricity (complex electrical energy consumption devices) in a household have average consumption of energy presented on Table 3.3.

The third group of electrical appliances is a group of multi-state appliances, such as washing machine and dishwasher. Duty cycle of these appliances can be described by a graph with different states. The identification of multi-state appliances is performed based on appropriate operating states sequence. The appliances with on, off and standby modes are 3-state appliances.

The last group includes appliances with continuously varying power. The identification of such appliances is the most difficult because these appliances cannot be described by well-defined states.

Table 3.1. Typical parameters of basic electrical energy consumption devices

Type of basic electric energy consumption device	Suppy volt-age [V]	Power [W]	Typical power [W]	Power factor $\cos(\varphi)$ for 50Hz	Inrush current factor $k_r = I_r/I_n$	Typical inrush time [ms] t_r	Current harmonics related to funda-mental
incandescent lamp (classic light bulb)	230	15 to 500	40 to 60	1	11	50	0
fluorescent lamp (fluorescent tube)	230	5 to 250	11 to 23	0.5 to 0.9	350	1	3h – 80% 5h – 50% 7,9,11,13 h – 30%
compact fluorescent lamp CFL (energy-saving light)	230	4 to 80	20 to 40	0.5 to 0.8	?	?	3h – 25% 5h – 8% 7h – 4% 9h – 2%
transformer power supply	230	2 to 300	100VA	0.93	55	20ms	2h > 20%
switching power supply without PFC	230	2 to 75	40VA	0.75	60	?	3h– 100% 5h – 90% 7h – 80% 9h – 70% 11h– 60% 17h,19h – 50%
switching power supply with PFC	230	75 to 300	150VA	0.8 to 0.99	?	?	?
electrical heaters	230	up to 3000	1500	1	1	0	0
microwave ovens	230	800 to 1600	1000	0.5 to 1	8	40	3h–25% 5h – 15% 7h – 5% 9h – 3%
electric motors (one phase)	230	Up to 2000	1000	0.6 to 0.8	15	1s	Approx. 0%
electric motors (3 phase)	400		3000	0.8	20	few s	Approx. 0%

Table 3.2. Summary of complex electrical energy consumption devices and their basic components

	incandescent lamp	fluorescent lamp	Heater	Induction motor	Commutator motor	Switching power supply	Transformer power supply	Three-phase motor
Aquarium (fluorescent lamps, heater, pump)		X	X	X				
Kettle(2l)			X					
Coffee Maker			X			X		
personal computer						X		
Microwave oven						X		
Electric cooker						X		
Refrigerator and freezer (300l)						X		
personal computer						X		
Microwave oven						X	X	
Electric cooker	X		X					
Refrigerator and freezer (300l)				X		X		
Blender / Food Processor					X			
Vacuum cleaner					X			
Washing machine (5kg)			X	X		X		
Instantaneous water heater			X					
Hair dryer / curling iron			X		X			
Fluorescent tube		X						
Compact fluorescent lamp		X						
Television		X				X		
Toaster		X						
Video / DVD						X		
Kitchen fan				X				
Freezer				X		X		
Hi-Fi						X		
Dishwasher			X			X		
Incandescent lamp 40W	X							
Incandescent lamp 60W	X							
Incandescent lamp 100W	X							
Iron			X					

Table 3.3. Average consumption of electrical energy in household according to one of distribution companies in Poland

Type of basic electric energy consumption device	Typical power [W]	Typical 24h working time		Typical	
		Hours	Minutes	Number of devices	24h load [kWh]
Aquarium	130	12	0	1	1.56
Kettle (2l)	2000	0	5	1	0.17
Coffee Maker	900	0	14	1	0.21
personal computer	300	2	0	1	0.6
Microwave oven	1000	0	10	1	0.17
Electric cooker	6000	0	24	1	1.08
Refrigerator and freezer (300l)	100	24	0	1	1.08
Blender / Food Processor	400	0	3	2	0.04
Vacuum cleaner	1500	0	3	1	0.08
Washing machine (5kg)	2000	0	27	1	0.9
Instantaneous water heater	5500	1	20	1	7.33
Hair dryer / curling iron	1200	0	8	1	0.16
Fluorescent tube	40	4	0	1	0.16
Compact fluorescent lamp	11	4	0	6	0.26
Television	100	3	0	2	0.6
Toaster	850	0	10	1	0.14
Video / DVD	40	1	4	1	0.07
Kitchen fan	150	0	30	1	0.08
Freezer	100	24	0	1	1.08
Hi-Fi	100	3	0	1	0.3
Dishwasher	2200	0	25	1	0.92
incandescent lamp (classic light bulb) 40W	40	4	0	8	1.28
incandescent lamp (classic light bulb) 60W	60	4	0	2	0.48
incandescent lamp (classic light bulb) 100W	60	4	0	2	1.2
Iron	1000	0	8	1	0.13
	Overall consumption				
	Per year 72229kWh	Per month 602.4kWh	Per week 140.6kWh	Per 24h 20.08k Wh	

3.4 Appliance identification methods

Methods of appliance identification depend on sampling frequency (Figure 3.3). Analysis of measurements with low sampling frequency enables only observation of macroscopic changes of electrical parameters. Having measurements with sampling frequency greater than 0.5 kHz, it is possible to analyze the variability of parameters in time and frequency domain. For identification based on high-frequency electrical noise, it is necessary to use high sampling frequency of electrical parameters [3].

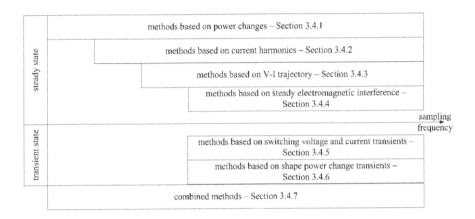

Figure 3.3 Classification of appliance identification methods based on sampling frequency

3.4.1   Identification methods based on power changes

The identification of appliances can be based on steady state analysis. During steady states, values of active power, reactive power, current or admittance is constant. The changes of electrical parameters correspond to changes in states of monitored appliances.

NIALMS system has been described for the first time by George Hart, who proposed identification of appliances based on changes in active and reactive power between steady states [2]. In this method measurements with sampling frequency of 1 Hz were used. This method allowed identification of appliances with a finite number of steady states with constant values of active and reactive power (Figure 3.4).

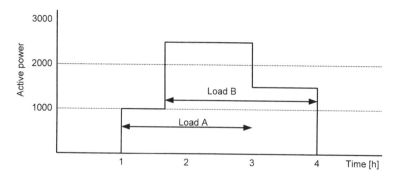

Figure 3.4 Changes of active power during operation of appliances

In the waveforms of active and reactive power the moments of changes are detected. The detected changes are plotted as points in the two-dimensional space of active and reactive power (Figure 3.5). The obtained points can be assigned to groups with approximately the same changes of active and reactive power. Groups with the same value but the opposite sign of change correspond to turning on and off of the same appliance.

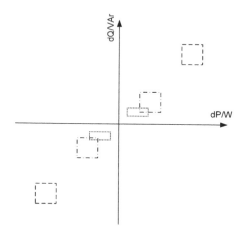

Figure 3.5 Groups of points corresponding to different appliances

For appliances with a finite number of steady states use of Finite State Machine (FSM) has been proposed (Figure 3.6) [2]. The method of building FSM models takes into account two constraints:

- steady states of appliances have different unique power values (UC-Unique Constraint)
- the sum of power changes throughout cycle of each appliance should be zero (ZLSC-Zero Loop Sum Constraint).

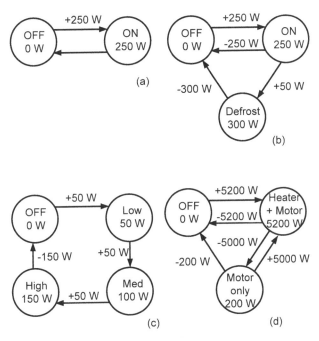

Figure 3.6 FSM models for different appliances

Figure 3.7 presents all steps of the algorithm proposed by Hart [2]. The first step is normalization based on the voltage measurement. Then in the normalized power waveform, power changes corresponding to appliance states changes are detected. Hart also introduced the division of the proposed method, depending on the degree of intrusiveness, on manual (MS-NIALM – Manual Setup NIALM) and automatic (AS-NIALM – Automatic Setup NIALM). In manual method several steps of the identification procedure can be omitted whereas in case of the automatic mode all steps, including preparation of appliance models, are necessary. The models building process is based on the analysis of the groups of points in a two-dimensional space of power changes and on the method of the FSM models building for multi-state appliances.

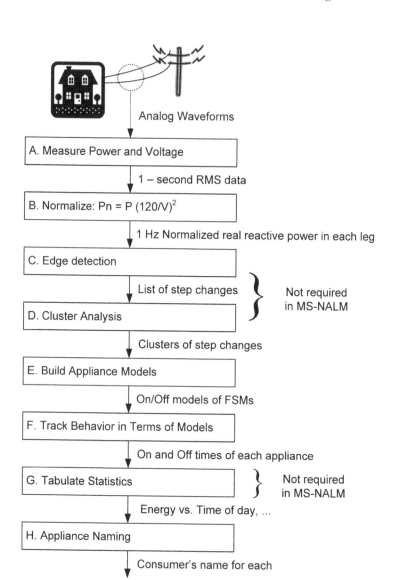

Figure 3.7 All the stages of identification method [2]

Several appliance identification methods based on changes in power have been presented in numerous publications [4–8]. However, mostly the proposed methods are modifications of the basic version of the algorithm proposed by Hart.

In the article by Baranski and Voss [6] a method in which data from the optical sensor directed to disk of induction meter was described. The disk rotation speed is read by detecting the mark on the disk. The proposed method consists of three steps (Figure 3.8). In the first step, power change events are grouped in term of repeated constant power. Having the set of all states, possible sequences are determined and for each sequence a hypothetical FSM model is obtained. In the next step a genetic algorithm is used to reduce number of FSM models. A constraint that one power change event can be assigned to only one FSM model is taken into consideration.

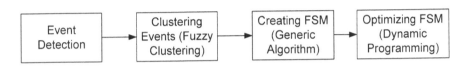

Figure 3.8 Successive stages of identification method [6]

A method to identify points on a two dimensional plane of power changes was presented in the article of Perez [5]. Each group of points of power changes can be specified with the vertices of the polygon. In this method, during identification of appliance, it is verified whether a point resulting from power change is inside the polygon. Therefore, the algorithm is called Point in Polygon.

In the next article [7], the use of temporal mining approach for appliance identification was proposed. This method is based on steady states analysis and detects repeated sequences of power changes. The aim of this method is to identify the shortest repeated sequences of power changes, summing to zero and corresponding to full operation cycle of each appliance.

Hidden Markov Models (HMM), based on measurements of electrical parameters may also be applied to identify appliances [9–12]. In this method steady states are determined on the basis of active power waveform as a function of time (Figure 3.10). The HMM model (Figure 3.10) is composed of steady states (circles), the likelihood of state changes (arrows) and the likelihood of remaining in the same state (directed arcs).

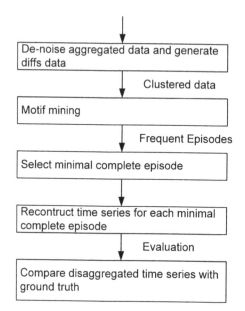

Figure 3.9 The temporal motif mining approach [7]

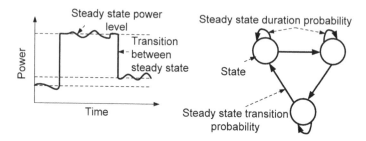

Figure 3.10 HMM as an appliance model [11]

The Factorial Hidden Markov Model (FHMM) is an extension of HMM that can be used to solve more complex problems with more than one independent time series (Figure 3.11) [10, 12]. Changes in active power are derived from aggregate time series and are assigned to different appliances. In addition, changes in appliance states are independent of each other and relatively rare because appliances mostly remain in the same state. The total power disaggregation may be directly reflected in the FHMM model. In the model there are independent chains of hidden variables, which describe the current status of each appliance. Changes of active power are expressed by observed variables. The active power changes are dependent on all hidden variables which determine current state of each appliance at the time.

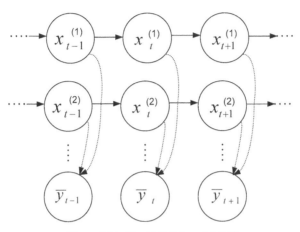

Figure 3.11 The FHMM model [10]

The use of FHMM model requires designation of the conditional probability of hidden states given observations. The most common approximation methods, used to solve this problem are: Gibbs sampling or variational methods. However, these methods often retain in local minima and are not an effective solution. By imposing some additional assumptions the problem can be simplified. Such effective method was described by Kolter and Jaakkola [10].

Parson and co-workers [9] proposed an iterative method with the use of HMM. The appliance identification in this case is performed using general appliance models. The general models include an estimate power values and transition probabilities. These models are adjusted based on aggregate measurements using expectation maximization (EM) algorithm. Then, the tuned models and modified Viterbi algorithm are applied to the appliance identification. This algorithm at each iteration identifies one appliance and filters out the impact of other appliances. Power assigned to identified appliance is subtracted from the aggregate measurements. On the basis of aggregate measurements reduced in previous iteration the next iteration is performed to identify the next appliance.

3.4.2 Identification methods based on current harmonics

In case of appliances with similar power change, they can be identified with the use of current harmonic analysis [13–20]. In the proposed methods the identification is based on information about active, reactive power change and current harmonics. The current harmonics can be calculated using measured data with 8 kHz sampling rate and phase-locked short time Fourier transform [13].

Having appliance signature containing information about current harmonics it is possible to use the Artificial Neural Networks (ANN) for appliance identification. The Figure 3.12 shows the structure of the ANN. In the input layer current harmonics values are provided. Whereas in the output layer the appliance identification is obtained. In this solution the ANN are used as nonlinear classifier.

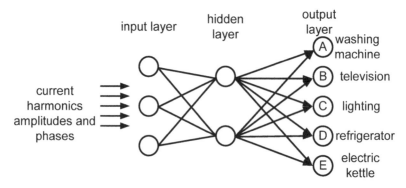

Figure 3.12 The proposed ANN structure

3.4.3 Identification based on V-I trajectory

Appliance identification can be performed using the signature in the form of the two-dimensional trajectory on the plane of normalized voltage and current [21, 22]. Based on the shape of the trajectory a set of parameters can be defined.

One of parameters that can be used for appliance identification is the shape of the average line of the trajectory (Figure 3.13-dashed line). For linear appliances the average line should be straight with an appropriate angle.

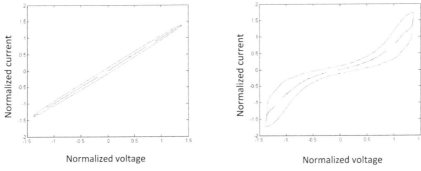

Figure 3.13 The shape of average line of the trajectory

Moreover, the trajectory can be dived into three segments. Information about the area of segments is specific for each appliance and can be used as a parameter during appliances identification. Figure 3.14 shows the relation between trajectory segments and current waveform for two different appliances.

Another parameter, that can be designated from the trajectory shape is the number of trajectory intersections. The number of intersections determine dominant current harmonics. In case of two intersections (Figure 3.15), in addition to fundamental current harmonic, the third harmonics is observed.

The following list contains all parameters describing the shape of the trajectory [21]:

1. Asymmetry
2. Direction
3. Area circled by the trajectory
4. Area of 3 segments
5. Shape of the average line
6. Number of intersection points
7. Angle of the middle segment
8. Maximum value in the middle segment

Having values of listed above parameters, the appliance identification can be performed, because these parameters are unique for different appliances and contain information about the electrical nature of the appliances. One of the proposed identification methods, in which trajectory shape parameters can be used is hierarchical clustering approach [22]. In this method the appliances are organized, based on the trajectory parameters, as a dendrogram which describes relationship between them.

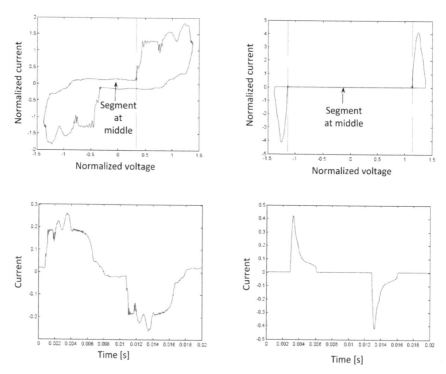

Figure 3.14 The division of the trajectory into 3 segments

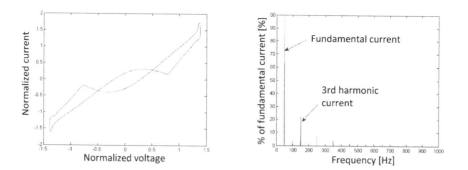

Figure 3.15 The intersections point of the trajectory

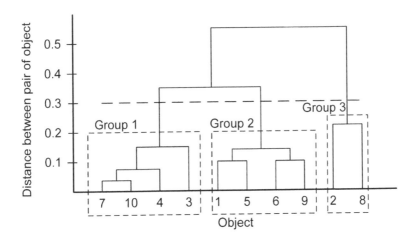

Figure 3.16 Appliances dendrogram in hierarchical clustering

3.4.4 Identification based on electromagnetic interference

In most electronic appliances Switched Mode Power Supplies (SMPS) are used. Such solution is more effective than classical transformer in terms of size, weight and cost. Moreover, in case of transformer the power is constant. Whereas electronic appliances with SMPS get current pulses containing a large amount of harmonics.

SMPS generate a high-frequency electrical noise (EMI-Electromagnetic Interference) during operation. The EMI noise is specific for different electronic appliances and can be used for the identification [23]. The noise can be divided into: continuous noise and switching noise. The switching noise occurs in short periods during state changes and will be discussed in the next section. The continuous noise is a narrowband with characteristic center frequency for different appliances. The frequency, for which the level of noise is the highest, is called switching frequency. For larger and smaller frequency than switching frequency the noise level decreases. The level of noise signal against frequency can be described with a normal distribution with defined mean and standard deviation values. For appliance identification, in this method, the use of switching frequency and normal distribution parameters was proposed. These parameters are different for each appliance and are saved in appliance library.

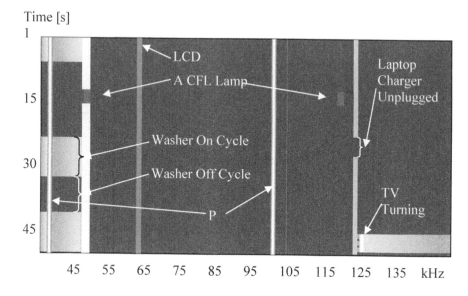

Figure 3.17 The frequency spectrogram [23]

The NIALM system that uses electromagnetic interferences for appliance identification was proposed by Gupta and co-workers [23]. Figure 3.18 shows a block diagram of the proposed system.

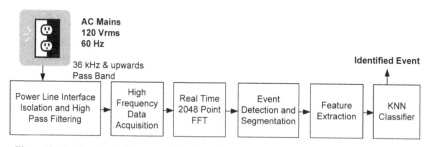

Figure 3.18 The block diagram of the identification based on continuous electromagnetic interferences

This system consists of Power Line Interface. The module provides an adequate filtration and isolation between the proposed system and electrical system. The next module used in this method is the data acquisition module. Based on data from the data acquisition module, the Fast Fourier

Transform is calculated. Then, the event detection module scans the signal for events when the noise level is greater than defined threshold value. The appliances are identified using k Nearest Neighbor and parameters saved in appliance library.

3.4.5 Identification based on switching voltage and current transients

Appliance identification can also be performed using noise emitted during transient states – turn on and off. During transient states high-frequency changes, called noise, in voltage and current may be observed. The noise is observed due to rapid short-circuits and openings. The reason for rapid changes are deflection and rebound of contacts and arcing. On the Figure 3.19 the rapid changes in voltage (switching voltage) during turning of a blender and continuous noise during steady state is presented.

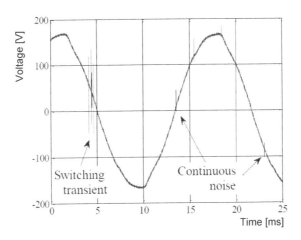

Figure 3.19 Switching voltage transients

The switching voltage are non-periodic and may be observed in a short period of time during transient states. Therefore, the use of Short Time Fourier Transform (STFT) and Continuous Wavelet Transform (CWT) were proposed for appliance identification [24] [25] [26]. In both cases, the transient waveforms are shown as a function of time and frequency. However, between STFT and CWT, the CWT was proved to be better solution due to better accuracy of identification and reduction of the dimension of feature vectors [24, 25]. Having the results of transforms, feature vectors are obtained. The feature vectors are unique and different for each appliance thus they can be used for the identification. Then for classification of obtained feature vectors to different groups of appliances Support Vector Machine (SVM) [24] and Artificial Neural Networks (ANN) [25] were tested.

3.4.6   Shape power change transient methods

The steady state identification methods may be improved by applying shape of parameters and other information about transient states. The use of transient power shape was proposed by Norford and co-worker [27]. In this method, the appliance identification is performed based on active, reactive power changes and the transient power shape as an additional feature. For transient power shape analysis, distance metric can be applied. However, the use of power shape is problematic because of poor repeatability.

Transient properties are connected with the physical task of the appliance and for example the turn-on transient energy may be applied in the appliance identification [28]. Chang applied data with 15 kHz sampling rate for appliance identification. In this identification method, with transient energy analysis, Genetic Programming (GP) was proposed for searching optimum feature vector, and Artificial Neural Network (ANN) for classification of appliances. The transient energy can improve accuracy of the identification [28].

3.4.7   Combined identification methods

In the end-user space commonly several 3-phase appliances can be found. The 3-phase appliance can be easily identified based on the analysis of electrical parameters changes in 3 circuits. In a usual household the number of 3-phase appliances is much smaller than 1-phase appliances. Therefore, it is simpler to identify several 3-phase appliances than many 1-phase appliances. After identification of all 3-phase appliances, the assigned power can be subtracted from aggregate power and then it is relatively easier to identify 1-phase appliances remaining in the household. To detect changes in 3 phases the use of voltage and current sensors with sampling rate of 1 Hz at the main switch was proposed [29, 30].

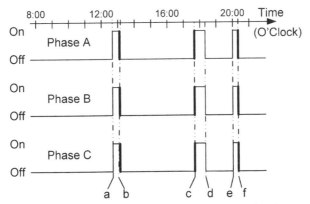

12:33.05 (a), 17:41.34 (c), 20:01.15 (e) – water heater on
13:12.28 (b), 18:21.04 (d), 20:19.15 (f) – water heater off

Figure 3.20 Identification of 3-phase appliances

Due to limitations of different identification methods, it was proposed to use several methods in parallel based on different signatures [30]. As the final result it is considered the most frequently obtained result from all used methods. Such Committee Decision Mechanism (CDM) is called Most Common Occurrence (MCO). Least Unified Residue (LUR) was also tested as CDM, in which the final result is the result with the least unified residue between unknown signature and the signature from appliance library. The full list of signatures used in different applied methods is as below:

1. Current Waveform (CW)
2. Eigenvalues (EIG)
3. Instantaneous admittance waveform (IAW)
4. Instantaneous power waveform (IPW)
5. Harmonics
6. PQ – power changes
7. Switching Transient Waveform (STW)

Trung and co-workers proposed to identify appliances using the following signatures [31]:

1. Shape type of changes
2. Electromagnetic interference (EMI)
3. Different power
4. Power factor
5. Harmonics

The listed above parameters are designated simultaneously. Then in the event detection step, three transient detection modules are used to detect transients of three basic appliance types: motor, lighting and heating. Having the results of the event detection step, the parameters can be compared with values saved in the library to identify the appliances.

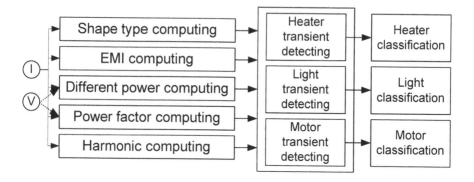

Figure 3.21 Combined method with 3 detection modules [30]

3.5 Comparison of different methods

In the disaggregation methods based on power changes it is possible to use data with low sampling rate – around 1 Hz. Low sampling rate reduces the requirements for hardware. The power change signature is additive. It means that when two appliances change state, the power change will be equal to sum of the two changes. In most methods with power change signature only appliances, for which the change of power exceeds a certain threshold, can be identified. In case of large number of appliances, significant number of group of points is obtained on the power change plane. That's why with increasing number of appliances, the methods based on power change gets more difficult. Moreover, it is not possible to identify appliances with continuously varying power, for which separate states with fixed power cannot be distinguished. Therefore, these methods may be applied to identify a limited number of appliances with unique power change values.

The steady state methods usually are performed in batch format, on the basis of data for a specified time. In case of some appliances, the stabilization after transient state can be significant. Due to batch format and significant stabilization time, the steady states methods can be used to monitor appliances in near real time.

To improve identification accuracy and to identify appliance with similar power change, analysis of current harmonics can be considered. However, in case of linear and quasi-linear appliances, the occurrence of current harmonics is limited.

It is also possible to distinguish appliances using transient or continuous electromagnetic interference analysis. The interferences are specific for different appliances, and depend on the relative position of measuring point and on the point of electrical outlet with connected appliance.

Due to the fact that different identification methods give better results for different groups of appliances, it seems beneficial to apply parallel methods. In this solution various methods operate in parallel on different signatures and the final result is based on results of all methods. Currently, in available literature there is lack of comparison of different methods on the same dataset.

The research area of non-intrusive monitoring systems and appliance identification methods is open and many questions still remain to be answered. Therefore, more work needs to be done to explore better this research field and improve methods that can be proposed for end-users.

Acknowledgement

This work has been accomplished within the project PBS2/A4/0/2013 "The Non-invasive System for Monitoring and Analysis of Electricity Consumption in the Area of the End-user" financially supported by the National Centre for Research and Development.

References

[1] S. Darby. The effectiveness of feedback on energy consumption: a review for DEFRA of the literature on metering, billing and direct displays., April 2006.

[2] G.W. Hart. Nonintrusive Appliance Load Monitoring. *Proceedings of the IEEE*:1870-1891, December 1992.

[3] M. Zeifman and K. Roth. Nonintrusive appliance load monitoring: Review and outlook. *IEEE Transactions on Consumer Electronics*:76-84, February 2011.

[4] A. Cole and A. Albicki. Algorithm for nonintrusive identification of residential appliances. *ISCAS '98 - Proceedings of the 1998 IEEE International Symposium on Circuits and Systems*, 3:338-341, 31 May - 3 June 1998.

[5] M. N. V. Perez. A Non-Intrusive Appliance Load Monitoring System For Identifying Kitchen Activities., 2011.

[6] M. Barański and J. Voss. Detecting patterns of appliances from total load data using a dynamic programming approach. *Data Mining, 2004. ICDM '04. Fourth IEEE International Conference*:327-330, November 2004.

[7] H. Shao, M. Marwah, and N. Ramakrishnan. A Temporal Motif Mining Approach to Unsupervised Energy Disaggregation. *1st International Workshop on Non-Intrusive Load Monitoring*, May 2012.

[8] A.J. Bijker, X. Xiaohua, and Z. Jiangfeng. Active power residential non-intrusive appliance load monitoring system. *AFRICON '09*:1-6, September 2009.

[9] O. Parson, S. Ghosh, M. Weak, and A. Rogers. Non-intrusive load monitoring using prior models of general appliance types. *Proceedings of theTwenty-Sixth Conference on Artificial Intelligence (AAAI-12)*:356-362, July 2012.

[10] J. Z. Kolter and T. Jaakkola. Approximate Inference in Additive Factorial HMMs with Application to Energy Disaggregation. *Proceedings of the Fifteenth Conference on Artificial Intelligence and Statistics*:1472-1482, April 2012.

[11] T. Zia, D. Bruckner, and A. Zaidi. A hidden Markov model based procedure for identifying household electric loads. *IECON 2011 - 37th Annual Conference on IEEE Industrial Electronics Society*:3218-3223, November 2011.

[12] Z. Ghahramani and M.I. Jordan. Factorial Hidden Markov Models. *Machine Learning*(29):245 - 273, November - December 1997.

[13] C. Laughman et al. Power Signature Analysis. *IEEE Power and Energy Magazine*:56-63, March-April 2003.

[14] D. Srinivasan, W. S. Ng, and A. C. Liew. Neural-Network-Based Signature Recognition for Harmonic Source Identification. *IEEE Transactions on Power Delivery*:398-405, January 2006.

[15] K. Yoshimoto, Y. Nakano, Y. Amano, and B. Kermanshaki. Non-Intrusive Appliances Load Monitoring System Using Neural Networks. *ASCEEE Summer Study on Energy Efficiency in Buildings*, 2000.

[16] H. Mori and S. Suga. Power system harmonics prediction with an artificial neural network. *IEEE International Sympoisum on Circuits and Systems*, 2:1129-1132, June 1991.

[17] R.S. Prasad and S. Semwal. A simplified new procedure for identification of appliances in smart meter applications. *2013 IEEE International Systems Conference (SysCon 2013) Proceedings*:339 - 344, April 2013.

[18] J. Lai, L. Suhuaui, and L. Jiaming. An Approach of Household Power Appliance Monitoring Based on Machine Learning. *Fifth International Conference on Intelligent Computation Technology and Automation (ICICTA)*:577 - 580, January 2012.

[19] J. Lei, L. Suhuai, and L. Jiaming. Automatic power load event detection and appliance classification based on power harmonic features in nonintrusive appliance load monitoring. *8th IEEE Conference on Industrial Electronics and Applications (ICIEA)*:1083 - 1088, June 2013.

[20] M. Akbar and D. Z. A. Khan. Modified Nonintrusive Appliance Load Monitoring For Nonlinear Devices. *IEEE International Multitopic Conference, INMIC*:1-5, December 2007.

[21] H. Y. Lam, K. H. Ting, W. K. Lee, and G. S. K. Fung. An Analytical Understanding on Voltage-Current Curve of Electrical Load. *Iternational Conference on Electrical Engineering (ICEE)*, 2006.

[22] H. Y. Lam, G. S. K. Fung, and W. K. Lee. A Novel Method to Construct Taxonomy of Electrical Appliances Based on Load Signatures. *Consumer Electronics, IEEE Transactions*:653-660, May 2007.

[23] S. Gupta, M. S. Reynolds, and S. N. Patel. ElectriSense: Single-Point Sensing Using EMI for Electrical Event Detection and Classification in the Home. *Proceedings of the 12th ACM International Conference on Ubiquitous Computing*:139-148, 2010.

[24] C. Duarte, P. Delmar, K. W. Goossen, and E. Gomez-Luna. Non-intrusive load monitoring based on switching voltage transients and wavelet transforms. *Future of Instrumentation International Workshop (FIIW), 2012*:1-4, October 2012.

[25] Y.-C. Su, K.-L. Lian, and H.-H. Chang. Feature Selection of Non-intrusive Load Monitoring System using STFT and Wavelet Transform. *e-Business Engineering (ICEBE), 2011 IEEE 8th International Conference*:293-298, October 2011.

[26] C. Hsueh-Hsien, C. Kun-Long, T. Yuan-Pin, and L. Wei-Jen. A new measurement method for power signatures of non-intrusive demand monitoring and load identification. *Industry Applications Society Annual Meeting (IAS), 2011 IEEE*:1-7, October 2011.

[27] L. K. Norford and S. B. Leeb. Non-intrusive electrical load monitoring in commercial buildings based on steady-state and transient load-detection algorithms. *Energy and Buildings*(24):51 -64, 1996.

[28] H.-H. Chang, C.-L. Lin, and J.-K. Lee. Load Identification in Nonintrusive Load Monitoring Using Steady-State and Turn-on Transient Energy Algorithms. *Proceedings of the 14th International Conference on Computer Supported Cooperative Work in Design*, 2010.

[29] J. Liang, S. Ng, G. Kendal, and J. W. M. Cheng. Load Signature Study - Part I: Basic Concept, Structure, and Methodology. *IEEE Transactions on Power Delivery*:551-560, April 2010.

[30] J. Liang, S. Ng, G. Kendall, and J. Cheng. Load Signature Study - Part II: Disaggregation Framework, Simulation, and Applications. *IEEE Transactions on Power Delivery*:561-569, April 2010.

[31] K. N. Trung et al. Using FPGA for real time power monitoring in a NIALM system. *International Symposium on Industrial Electronics (ISIE)*:1-6, May 2013.

Biographies

Wiesław Winiecki, professor of Measurement Science at the Institute of Radioelectronics, Warsaw University of Technology, has thirty seven-year research experience in the field of measurement and control systems, including the development and implementation of various kinds of measurement devices and systems. The record of his achievements in this respect comprises more than 200 research publications, 2 monographs, 1 book and 1 academic textbook, as well as over 100 reports on scientific research and implementation. He is a member of the Committee on Metrology and Scientific Instrumentation of the Polish Academy of Sciences and the Vice-President of the Polish Society for Measurement, Automatic Control and Robotics POLSPAR.

Krzysztof Liszewski graduated from the Faculty of Electronics and Information Technology of the Warsaw University of Technology. He has worked as a wind energy analyst for 3 years. He is also a member of the research team at Warsaw University of Technology. He is interested in database systems, SCADA systems, renewable energy, monitoring and optimization energy consumption systems.

4

Design and testing of an electronic nose sensitive to the aroma of truffles

Despina Zampioglou and John Kalomiros

Department of Informatics Engineering, Technological and Educational Institute of Central Macedonia, Serres, Greece

Abstract

An "electronic nose" based on a low-cost array of gas-sensors is described. It is designed for the detection of Volatile Organic Compounds (VOCs) emanated from samples of the ascomecyte Tuber or truffle. These fungi have highly appreciated gastronomical and nutritive merits and they own a variable characteristic aroma depending on their stage of maturation and place of origin. The results show that an intelligent odor-discriminating system based on a gas sensor array can contribute to the identification and classification of truffles.

Keywords: electronic nose; olfactory systems; gas sensors; data acquisition; truffle aroma

4.1 Introduction

4.1.1   Overview of e-nose systems

Odor is commonly used to assess quality in the food, beverage and perfume industry. Human sensory analysis is preferably used but being highly subjective it is often assisted by objective tools, like mass spectrometry and air chromatography. Such procedures require expensive non-portable equipment and do not usually accommodate real-time performance [1]. In many applications electronic nose systems have been developed and tested. Such systems are based on gas-sensor arrays that produce a characteristic

V. Haasz and K. Madani (Eds.), Advanced Data Acquisition and Intelligent Data Processing, 59–81.

pattern of each odor under test. First the system is trained using a set of test samples and then pattern recognition software can identify or classify the product under test [2, 3].

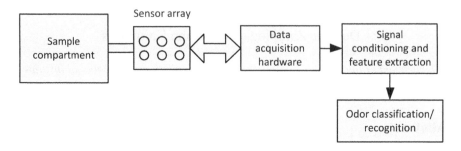

Figure 4.1 The stages of an electronic olfactory system

In the human olfactory system each single odor is perceived as a mixture of volatile organic compounds (VOCs). Odor molecules reach the nostrils and interact with receptor neurons sensitive to a wide range of compounds. Information is passed to the central nervous system in the brain where signals are separated, patterns are identified and odors are discriminated and classified. Mimicking this function, electronic noses usually consist of multi-sensor arrays, each sensor having a different sensitivity profile over the range of compounds expected in the target samples. An important requirement is the interaction of odor molecules with the sensor material to be fast and reversible. It is necessary that the sensor baseline drift is negligible and calibration to the initial state is replicable. Power consumption, size and sensor cost are also important considerations for portable systems [3].

Figure 4.1 illustrates the stages of an electronic olfactory system. The sample compartment is the space where volatile compounds are accumulated. The sensor array receives the gas flux and produces a response which is then preprocessed and conditioned in order to compensate for noise, varying temperature, changing relative humidity or baseline drift over time. An appropriate feature vector is extracted from sensor responses and is used in the last stage of the system for odor classification and identification.

Gas sensors commonly employed in e-nose systems are solid-state devices incorporating a sensitive layer in the form of thin or thick film. Gas molecules interact reversibly with the surface of the sensor material, by adsorption or absorption. Interaction with gas causes changes in some physical property of the reagent which then is used as gas detection principle. Different classes of sensor devices are based on different types of physical change [4–6]. Basic sensor classes include:

a. *Semiconductor devices*. Their ability to detect gas is based on the change of electrical conductivity. This class of sensors includes conducting polymer sensors (CPS) and metal oxide sensors (MOS) [3, 4, 5]. Tin oxide gas sensors are among the most prominent in e-nose systems. These sensors incorporate a heater since their sensitivity increases with operating temperature.

b. *Piezoelectric sensors*. Their operation is based on the change of resonant frequency of a piezoelectric substrate or bulk crystal, covered by a gas-sensitive coating [1, 5, 7]. Upon exposure to VOCs the mass of the reacting coating changes, which in turn results in a change in the frequency of the acoustic wave. This class of sensors includes surface acoustic wave (SAW) devices [7] and quartz crystal microbalance (QCM) devices [8, 9].

c. *Field-effect sensors*, like MOSFET. Their principle is based on the change of the work function of the gate and of the silicon dioxide layer, where the gate is deposited. The changes occur due to polarization effects upon interaction with gas molecules [5, 10]. Changes in the threshold voltage or in the source-drain current are used as sensor response properties.

d. *Optical sensors*. They consist of optical fiber coated with a polymer impregnated with a fluorescent dye. Upon interaction with gas the optical properties of the dye change resulting in a change of the optical response. The properties of the polymer control the response of the sensor to the target gases.

Sensor outputs in e-nose arrays are multiplexed, sampled and digitized using a data acquisition system, often based on a microcontroller or commodity data-acquisition hardware. The signals are denoised in a preprocessing step, using simple filtering techniques, like a moving average finite impulse response [11]. Baseline manipulation is also applied in this step. The signal versus time response of the various sensors in the multi-sensor array is used in order to extract a feature vector. Transient and steady state sensor responses can both be used for feature extraction; however the steady state signal response is often used because of its robustness. Principal component analysis (PCA) is commonly applied in order to reduce the vector dimensionality. Classification algorithms, like k-nearest neighbors (k-NN), artificial neural networks (ANN), radial basis function (RBF) and support vector machines (SVM) are used with training sets of odorous samples and their classification accuracy is validated using a set of validation samples. Subsequently, the trained system can classify input test odors to one of the pattern classes under consideration [12]. The classification step can be executed in an embedded processor or in an interfacing computer running a host application.

Several commercially available electronic noses have appeared in recent years, some using arrays of one type of sensors, others combining two or several types [13]. A number of applications have been explored for various industries, including food [8], environmental, biomedical [14], automotive and security applications. A survey of applications and advances is presented in [15]. Real-time portable electronic nose systems are also being researched since they are suitable for environmental monitoring. They can be applied to mobile robots in order to assist in search and rescue operations in toxic environment or in hazardous situations [16].

4.1.2   Application of electronic noses to food products

E-nose systems have been applied widely to the recognition and classification of natural products and especially to the assessment or discrimination of various qualities of consumable products [8]. Such systems are finding their way to industrial applications for purposes like food safety by identifying contaminants, biochemical composition by identifying the basic constituents and monitoring the effects of food treatment and processing [17]. Studies based on electronic noses have been directed mainly to the determination of meat and fish freshness [11, 18], the effects of meat processing [19], detection of fruit and vegetable aroma and ripeness [20–22], wine maturity and winery process assessment [23], and discrimination between various blends of coffee [24]. Potential applications in brewery [25] have also been considered. The off-odors of grains were used for quality classification [26] and various cheeses were classified according to their types and maturity stages [27, 28]. One particular study focuses on measuring volatiles from mushroom species [29] and is related to our present study on detecting and identifying the aroma of truffle specimens.

The most common olfactory sensors used in the food industry are metal oxide (MOS) and conducting polymer (CP) sensors. Laboratory-made as well as commercial systems have been used for food assessment and classification. The trend is to use application-specific systems based on selected sensors sensitive to the target volatiles. There is also a continuing quest for small-size, lightweight, portable systems, based on low-power microprocessors or microcontrollers, able to assess and classify products in the field. Our present study is towards this direction.

4.1.3   Contribution and organization of the present work.

In this work we explore the feasibility to implement an electronic nose system sensitive to the aroma of ascomycete Tuber, which is commonly referred to as truffle. Truffles are species of an underground mushroom that grows at the roots of some trees or bushes. They are highly appreciated for their gastronomical and nutritive merits. As they do not have above ground organs, they are very difficult to discover in nature and in most cases this happens randomly. The most remarkable among indications used for the location of truffles is the intense smell they emanate in the period of their maturity. Truffles synthesize an enormous diversity of volatile organic compounds resulting in their characteristic aroma. A lot of research has been focused on the composition of VOCs related to truffle aroma [30]. The present study is directed to the development of a low-cost, MOS based and microcontroller-based electronic nose, able to respond to the presence of truffle specimens. An array of six different metal-oxide gas sensors is tested against the truffle volatiles. The response to a variety of truffle specimens is monitored. A feature vector is extracted based on the transient and steady-state response of the MOS sensors. The system can basically support truffle detection and identification in laboratory conditions. Principal component analysis (PCA) indicates that our sensor array can discriminate truffle aroma from other fruity odors such as mature apple, as well as alcohol, to which most of the sensors typically respond. It also tends to discriminate between different truffle species, like *Tuber borchii*, *Tuber brumale*, *Tuber magnatum* and *Tuber macrosporum*. Artificially scented oil and conserved truffles have been tested as well. Future research is needed to refine the system and test its ability to further classify truffle specimens according to their multitude species, maturity and probably their place of origin, since aroma variability in truffles has been attributed to these factors [30]. The development of an olfactory standard for the quality of truffle specimens is of great interest for the industry.

The rest of the chapter is organized as follows. In Section 4.2 basic information is given about the diverse species of ascomycete Tuber and their variable organic volatile compounds. Understanding the nature of the target molecules is important to the selection of specific gas sensors. In Section 4.3 the multi-sensor array is described in detail along with the sample compartment/sniffing system and the microcontroller-based data acquisition, transmission and signal preprocessing system. In Section 4.4 experimental results from different truffle specimens are presented. In Section 4.5 the feature extraction process is described along with the pattern identification analysis. Section 4.6 concludes the paper.

4.2 Ascomycete tuber and its volatile compounds

4.2.1 Truffle species

Truffles are a relatively rare species of subterranean fungi typically found near mycorrhizal roots of woody plants in shrub areas, groves or forests. The symbiotic species are usually oaks, willows, pines and hazel trees. The gastronomical and nutritive merits of truffles make these fleshy fungi one of the most exquisite dishes worldwide. Truffles are native mainly to temperate regions and occur in diverse families and genera. Edible truffles mainly belong to the large group of fungi called ascomycetes, to the family Tuberaceae and to the genera Tuber and Terfezia. Ascomycetes or sac fungi are characterized by a saclike structure, the ascus, which contains six to eight ascospores in the sexual stage. The Tuber genus includes about one hundred eighty five species.

Figure 4.2 *Tuber brumale* (left) and *Tuber borchii* in their forest beds.
Cross sections are also shown

The most valued among them are the winter black truffle *Tuber melanosporum* which is found throughout the Northern hemisphere and the winter white truffle *Tuber magnatum* which is native to Southern Europe. Similar to the melanosporum are *Tuber brumale*, which is a musky black winter truffle and *Tuber aestivum*, which flourishes mainly during the summer and can be found throughout Europe. Other common truffles species include *Tuber borchii*, which is a white-beige or brown mushroom and *Tuber uncinatum* which ripens from October through January and is the autumn variety of *Tuber aestivum*.

Additionally to their culinary merits truffles are also believed to have healing properties against muscular pains and arthritis and to lower the cholesterol levels. They have also been thought to have powerful aphrodisiac properties [31]. Images of truffle fungi are shown in Figure 4.2.

In order to determine the locations where truffles grow, various indications are used, as for example the specific trees of the region, the total lack of grass, swarms of yellow flies flying in low altitude over the truffle areas, the light elevation of the soil and the clefts of the ground under which this mushroom grows. But the most remarkable indication of all is that when truffles mature, they emanate an intense smell that can be detected from a long distance by some animals like pigs, squirrels, deer, dogs and bears. During the maturity period, each genus of truffle emits its own smell.

4.2.2 Key truffle volatiles

Fungi synthesize a diversity of volatile organic compounds. More than 200 VOCs have been described from the fruiting-body of natural truffles in the last twenty years [30, 32–34]. Organic volatiles are identified mostly by gas chromatography/mass spectrometry (GC/MS), using solid phase microextraction (SPME) for sampling [33]. A few of them are considered as key to the communication of truffles with other organisms.

The fresh truffles contain a number of organic molecules, like alcohols, aldehydes and ketones. The smell of the truffle is actually due to the molecule dimethyl sulfide or CH_3SCH_3 [35], and to other sulfur-VOCs, like $CH_3CH_2CH_2SCH_3$ and $CH_3CH=CHSCH_3$ [36]. Other truffle compounds are eight-carbon-containing volatiles (C8-VOCs), predominantly 1-octen-3-ol, that may act as signals to plants [37] and androstenol [38], which is an animal pheromone. The relative quantity of alcohols to aldehydes and ketones varies, but all genotypes contain dimethyl sulfide, which acts as an attractant to animals [35]. A single truffle fruiting body typically contains 20-50 volatile compounds, and the composition of these volatiles might depend on genotypic variability, geographical location and/or maturation stage [30, 39]. When the truffles are kept over a period of time the volatile sulfur compounds escape faster than the other molecules and their release into the air gives these fungi their strong pungent smell.

4.3 Design of the e-nose system

4.3.1   Sensor array

An array of different gas sensors, each one able to capture different volatile compounds is used in the preliminary stage. A gas sensor array permits to improve the selectivity of a single gas sensor, and shows the ability to classify different odors and quantify component concentration [12]. A trial system has been designed and implemented based on low-cost solid state MOS sensors. Their sensing element is metal oxide, most typically tin dioxide (SnO_2). In the presence of detectable volatile compounds the sensor's conductivity increases, depending on the gas concentration in the air. The sensors use a heater in order to increase oxygen adsorption at grain boundaries. Oxygen forms potential barriers which control the sensor's resistance. In the presence of deoxidizing gases, like organic volatiles, potential barriers fall and the sensor conductivity is increased [11].

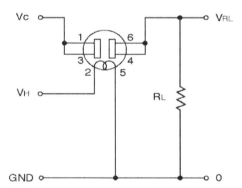

Figure 4.3 The basic measuring circuit. A heater voltage and a measurement bias voltage are required

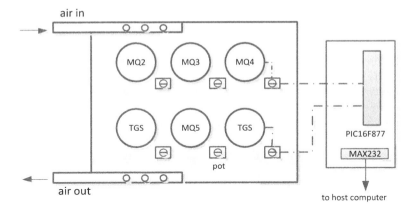

Figure 4.4 Diagram of the six MOS sensors array, with calibration potentiometers. On the right, the data-acquisition system is shown

Each sensor resistance is measured using a simple measurement circuit formed with a load resistor R_L and a buffer. For calibration purposes R_L is partly formed by a potentiometer. The load resistance can also be varied to adjust the sensor sensitivity to a particular volatile compound. Figure 4.3 presents a typical sensor diagram, where the heater, the sensor electrodes and the load resistance are shown.

Table 4.1 Elements of the tested sensor array

Sensor	Type/Manufacturer	Main detectable gas
TGS 822	MOS/ Figaro Eng. Inc.	Organic solvent vapors, Acetone, ethanol
TGS 2602	MOS/ Figaro Eng. Inc	VOCs, odorous gases
MQ2	MOS/ Hanwei Electronics	General combustible gases-LPG
MQ3	MOS/ Hanwei Electronics	Alcohol
MQ4	MOS/ Hanwei Electronics	CH_4, natural gas
MQ5	MOS/ Hanwei Electronics	LPG, natural gas, town gas

In our present implementation six solid-state gas sensors are used. Five sensors are from the MQ series manufactured by Hanwei Electronics and two from the TGS series manufactured by Figaro Engineering Inc. They are presented in Table 4.1, along with their corresponding main target gases. However, all sensors are sensitive to a variety of gases, according to their data sheets. A diagram of the sensor array is presented in Figure 4.4.

4.3.2   Sample compartment and sensor chamber

The sensor array is sealed in the measurement compartment, where volatiles are injected through air-tubes. A diagram of the system, as it is configured at present, is shown in Figure 4.5. The samples to be measured are placed in the sample compartment where gas fumes are accumulated. At the beginning of each measurement the air pump is turned on and air is pumped out of the sensor compartment. Then valve 1 is turned to allow gas from the sample compartment to fill the sensor chamber and the air pump is turned off. The measurements reach a steady state in about one minute and a half and the sample is measured for a total of about 3 minutes. Following the sample measurement the air pump is turned on again and a "cleaning" step takes place by injecting dry air in the measurement compartment. A period of 10 minutes is allowed for the sensors to settle to their initial values. A sensor calibration step is also conducted periodically to ensure that all measurements have the same reference in dry air.

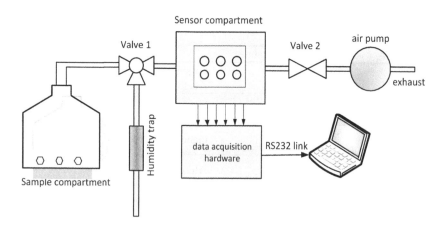

Figure 4.5 The configuration of the e-nose system

4.3.3   Data acquisition and processing system

Data acquisition is implemented with a PIC16F877 microcontroller by Microchip Corporation, using a six-channel on-chip analog to digital converter with 10-bit analysis (see also Figure 4.4). Depending on the noise present in the measurements, the analysis can be reduced to 8-bits. The measurement dynamic range can be adjusted by the potentiometers and in our experiments it is 2.0-5.0 Volts, the lower value corresponding to dry air. Sampling rate is 50 Hz and total measurement cycle is approximately 15 minutes. An RS232 serial interface is used for communication with a host computer. Since sensor response is not very fast, serial communication is adequate for data transmission. A LabVIEW™ host application performs data analysis and displays the results. Signal conditioning, consisting mainly of low-pass filters, is implemented in software and is used for noise suppression. A screenshot of the front panel is shown in Figure 4.6.

4.4 Sensor-array response to truffle aroma

Samples of several *Tuber* species were measured in the present study. The measured species appear in Table 4.2, along with information on truffle characteristics and maturity periods. All samples were collected from

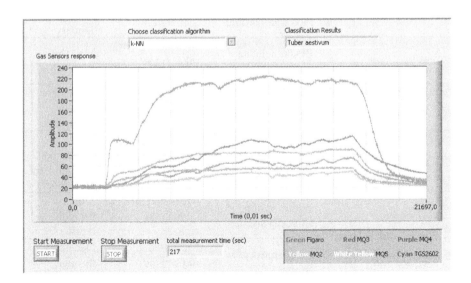

Figure 4.6 The front panel of the host LabVIEW™ data acquisition application

forests in eastern Macedonia, Greece, by a specialized cultivator and truffle hunter. Most truffle specimens were gathered during maturity; however the measured samples of *Tuber magnatum* are probably in their early stage of maturation and are relatively small. This truffle was discovered in Greece only recently [40] and is considered of great value since it is generally difficult to find. The bodies of the truffles of different kinds have different colors and textures and their aroma has some characteristic differences. Some of the samples were kept in a freezer for five days before being measured, as indicated in Table 4.2. After defrost, the fruiting body of the truffle was put into the sample compartment for a few minutes, according to the procedure described in Section 4.3. Typical results are shown in Figure 4.7. The rising part of the curves corresponds to the injection of the truffle fumes into the measurement compartment, while the falling part is monitored at the cleaning step. The most intense responses are given by sensors TGS2602 and MQ3 which are designed to detect general VOCs and alcohol, respectively. TGS822 which is a general-purpose gas sensor

Table 4.2 Measured truffle species and their characteristics

Truffle species	Color of fruiting body and other characteristics	Measurement conditions	Maturity period
Tuber uncinatum	Black, pronounced flavor and aroma	Measured as gathered	October through January
Tuber aestivum	Black, delicate aroma	Measured after 5 days in the freezer	May through September
Tuber macrosporum	Black or reddish brown	Measured as gathered	September through December
Tuber magnatum	White, very rare, methane-garlic aroma	Measured as gathered	Late September through late November
Tuber borchii	White, strong alliaceus aroma	Measured after 5 days in the freezer	January through April
Tuber brumale	Black, musky aroma	Measured after 5 days in the freezer	December through March

and MQ5, which is sensitive to natural gas, also give strong responses in most of our experiments. MQ4 and MQ2 also respond in all experiments, usually giving lower steady state values. The measured quantity is the resistivity of the metal-oxide sensor, translated to quantum states in the y-axis of Figures 4.7, 4.8 and 4.9. The patterns generated by the VOCs sensors in most other measurements are similar to those described above and differ only in details.

Beside the wild truffle specimens, the sensor-array response to a conserved truffle product was monitored. The odor of small truffle slices pickled in artificially scented olive oil was measured. Typical results are shown in Figures 4.9. The sensors respond in a way similar to that of Figure 4.7, however the measured values are lower. Again TGS2602 and MQ3 give the higher response, with TGS822, MQ5 and MQ2 following, with lower responses. Finally, sensor MQ4 is rather insensitive to the volatile compounds emitted by the artificially scented truffle samples.

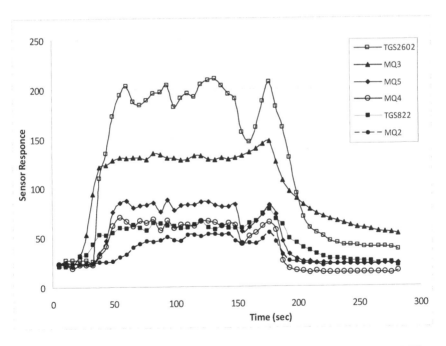

Figure 4.7 Response to *Tuber brumale*. Curves from top to bottom are from TGS2602, MQ3, MQ5, MQ4, TGS822 and MQ2 respectively

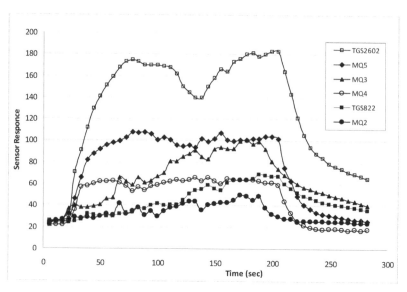

Figure 4.8 Response to *Tuber aestivum*. Curves from top to bottom are from TGS2602, MQ5, MQ3, MQ4, TGS822 and MQ2 respectively

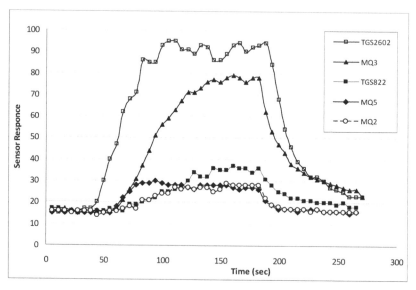

Figure 4.9 Typical response to conserved truffle specimens pickled in artificially scented oil. Curves from top to bottom are from TGS2602, MQ3, TGS822, MQ5 and MQ2, respectively

4.5 Discrimination of sensor patterns

4.5.1 Feature extraction from the measurement data

In Figure 4.10 the basic measurement characteristics of a typical sensor response to the fumes of truffle specimens are shown. Several attributes of the curves are used for feature extraction in the present study. The procedure is as follows:

(1) The baseline $x(0)$ is subtracted from the sensor response $x_s(t)$ in order to compensate for noise or drift, $y_s(t)=x_s(t)-x(0)$.
(2) The steady state value st is extracted from the upper part of the curve.
(3) The gradient $grad_1$ of the rising edge of the measurement ("on" derivative) is obtained, by calculating $\Delta y_1/\Delta t_1$, where Δy_1 is the 70% of the steady state value st and the corresponding time interval Δt_1.
(4) The gradient $grad_2$ of the falling edge ("off" derivative) is obtained, by calculating $\Delta y_2/\Delta t_2$, where Δy_2 is $(1-0,7)st$ and Δt_2 the time interval of the corresponding fall of the response.

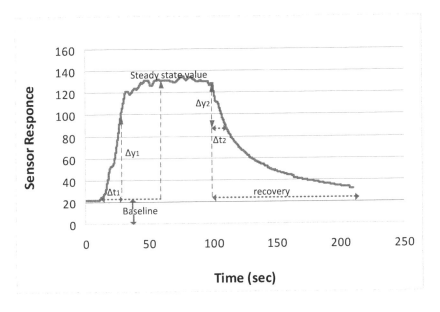

Figure 4.10 Typical sensor response to truffle VOCs and quantities used for feature extraction

The above parameters *st*, *grad₁* and *grad₂* are extracted from each sensor measurement. The total of six sensors results in an eighteen-dimensional feature vector. The features are rescaled using appropriate scale factors. Alternatively, one can choose only the steady state *st* of the sensor response and form a six-dimensional vector. Although both methods achieved some level of classification and recognition of the truffle samples, the gradient-enriched vectors generally produce better results.

4.5.2   Principal components analysis and k-NN

It is well known that PCA is an unsupervised linear feature extractor that determines an appropriate subspace of reduced dimensionality m in the original feature space of dimensionality d ($m \leq d$). For this purpose, PCA uses the most expressive features by computing the m eigenvectors with the largest eigenvalues of the $d \times d$ covariance matrix of n d-dimensional feature vectors [41]. PCA computes the most meaningful basis to re-express the data-set and in this way it reveals hidden data dynamics, while it suppresses noise and reduces redundancy in the measurement data. In our feature space we have defined eighteen dimensions, as described in the previous paragraph. In the following, the three or two principal components are extracted and plotted, while the possible clustering of data according to their corresponding species is examined. Principal eigenvectors are extracted using Singular Value Decomposition (SVD) [42].

In a first experiment we try to find the variance between truffle samples of various species and other fruity samples, like mature apple. Ethanol is also added to the measurement set, since most of the sensors used in our sensor array are sensitive to alcohol. The resulting PCA plot is shown in Figure 4.11. Truffles can basically be discriminated from other samples, especially ethanol, while truffle specimens present a variance themselves within the feature space. This variance is presented in the following plots and is associated with the measured Tuber species.

In Figure 4.12 PCA is applied to measurements of three different Tuber species, namely *Tuber magnatum*, gathered at an early stage of maturation, *Tuber uncinatum* and *Tuber macrosporum*. All these specimens are measured fresh as gathered, within three days from the day they were collected. The measurements of the mature apple specimens are also included for comparison. The relative contribution of the three principal components is 60%, 30% and 10% respectively. This relative contribution is typical for the presented plots.

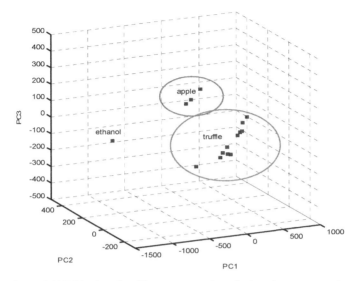

Figure 4.11 PCA plot of truffle, mature apple and ethanol measurements

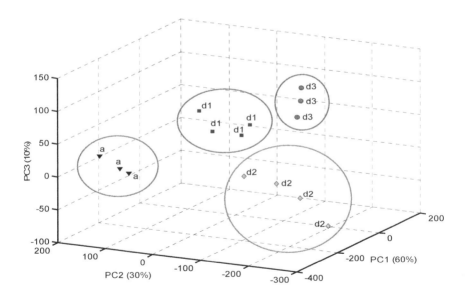

Figure 4.12 Projection of three principal components. d1 (squares): early *Tuber magnatum*. d2 (rhombs): *Tuber uncinatum*. d3 (circles): *Tuber macrosporum*. a (triangles): apple specimens

In Figure 4.13 a new Tuber species is added to the plot, namely *Tuber borchii*. Before these samples were measured they were kept in a freezer for five days. A degree of overlapping between measurements is visible. Overlapping is more pronounced as more measurements from different species become available. However, *Tuber borchii* can be discriminated from samples of *Tuber macrosporum* and *Tuber brumale* in the 2-D plot of Figure 4.14. Finally, a 3-D PCA plot derived from six-dimensional feature vectors, based only on the steady-state *st* of the sensor responses is shown in Figure 4.15. Although measurements now overlap more closely and intra-species variance is increased, coarse groups can still be formed in the feature space. Species clustering with 12 or 18 feature dimensions is usually more pronounced.

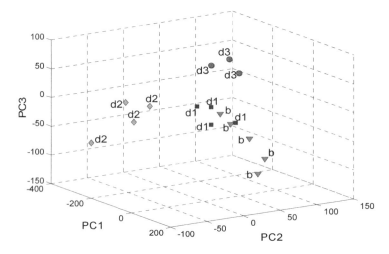

Figure 4.13 PCA plot derived from measurements of four different species. d1 (squares): early *Tuber magnatum*. d2 (rhombs): *Tuber uncinatum*. d3 (circles): *Tuber macrosporum*. b (triangles): *Tuber borchii*

A k-Nearest Neighbors (k-NN) classifier has also been trained and applied to the detection of truffle specimens in the laboratory. As a result, k-NN has a success rate of about 85% in identifying truffle samples as opposed to other odors to which the sensors respond, like mature apple, ethanol, oranges and ethereal oils. Artificially scented truffle products can also be discriminated from fresh specimens. Such tests have not yet been implemented in the field, where truffles are cultivated or grow spontaneously.

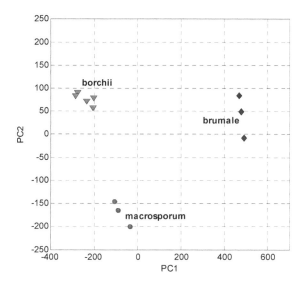

Figure 4.14 Plot of two principal components of measurements taken from *Tuber brumale*, *Tuber borchii* and *Tuber macrosporum* samples

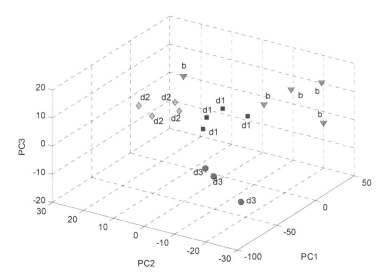

Figure 4.15 PCA plot with six-dimensional feature vectors based on the steady state value of each sensor response. Symbols are as in Figure 4.13

4.6 Conclusion

An electronic nose system based on a MOS sensor array is developed and tested against the volatile organic compounds emitted by truffle specimens. All the tested sensors are responsive to truffle aroma with the best response given by VOCs, alcohol and natural gas sensors TGS2602, MQ3 and MQ5, respectively. The presented results encourage that a low-cost electronic nose system can be developed and contribute to the investigation and identification of ascomycete Tuber or truffle. Such a device can be used for truffle quality and maturity control, both in the field and in the packing industry. The present study contributes to an already mature literature on intelligent applications of electronic olfactory systems.

References

[1]   H.T. Nagle, R. Gutierrez-Osuna and S.S. Schiffman, "The how and why of electronic noses" IEEE Spectrum, Vol. 35, No. 9 (1998) pp. 22-31.

[2]   "Handbook of Machine Olfaction: Electronic Nose Technology", T.C. Pears, S.S. Schiffman, H.T. Nagle, and J.W. Gardner (Eds.), Wiley-VCH, Weinheim, 2003.

[3]   K. Arshak, E. Moore, G.M. Lyons, J. Harris and S. Clifford, "A review of gas sensors employed in electronic nose applications", Sensor Review, Vol. 24, No. 2 (2004), pp. 181-198.

[4]   M. Griffin, "Electronic Noses: Multi-Sensor Arrays", Davidson College, NC 28036, USA (http://www.coffeeresearch.org/science/nose.pdf)

[5]   K.J Albert, N.S.Lewis, C.L. Schauer, G.A. Sotzing, S.E. Stitzel, T.P. Vaid, and D.R. Walt, "Cross-Reactive Chemical Sensor Arrays", Chem. Rev. Vol. 100 (2000), pp. 2595-2626.

[6]   S. Capone, A. Forleo, L. Francioso, R. Rella, P. Siciliano, J. Spadavecchia, D. S. Presicce, and A.M. Taurino, "Solid state gas sensors: state of the art and future activities", Journal of Optoelectronics and Advanced Materials, Vol. 5, No. 5 (2003), pp. 1335-1348.

[7]   Z.P. Khlebarov, A.I. Stoyanova, and D.I. Topalova, "Surface acoustic waves gas sensors", Sensors and actuators B: Chemical, Vol. 8 No. 1 (1992), pp. 33-40.

[8]   E. Schaller, J.O. Bosset, and F. Escher, "Electronic Noses and their application to food", Lebensmittel-Wissenschaft und Technologie, Vol. 31, No.4 (1998), pp. 305-316.

[9]   Y.H. Kim and K.J. Choi, "Fabrication and application of an activated carbon-coated quartz crystal sensor", Sensors and Actuators B, Vol. 87 (2002), pp. 196-200.

[10] E-L. Kalman, A. Lofvendahl, F. Winquist, and I. Lundstrom, "Classification of complex gas mixtures from automotive leather using an electronic nose", Analytica Chimica Acta, Vol. 403 No. 1 (2000), pp. 31-38.

[11] N. ul Hasan, N. Ejaz, W. Ejaz, and H. Seok Kim, "Meat and fish freshness inspection system based on odor sensing", Sensors 2012, 12, 15542-15557.

[12] R. Gutierrez-Osuna, "Pattern Analysis for Machine Olfaction: A Review", IEEE Sensors Journal, Vol. 2, No 3 (2002), pp. 189-202.

[13] E. Vannests, H. Geise, in "Handbook of Machine Olfaction, Electronic Nose Technology", (2003), ch. 7, Wiley-VCH, Weinheim, pp. 161-179.

[14] A. Wilson and M. Baietto, "Advances in electronic-nose technologies developed for biomedical applications", Sensors 2011, 11, 1105-1176.

[15] A.D. Wilson, M. Baietto, "Advances in Electronic nose technologies", Sensors 2009, 5099-5148.

[16] A. Holloway, G. Chliveros and A. Nabok "Real-time machine olfaction for mobile robot applications", In Proc. of IARP/EURON Workshop on Robotics for Risky Interventions and Environmental Surveillance (RISE08), Benicàssim (Spain), 2008.

[17] C. Di Natale, R. Paolesse and A. D'Amico, "Food and Beverage Quality Assurance", in Handbook of Olfaction: Electronic Nose Technology, T.C. Pears, S.S. Schiffman, H.T. Nagle, and J.W. Gardner (Eds.), Chapter 21, Wiley-VCH, Weinheim, 2003.

[18] N. El Barbi, E. Llobet, N. El Bari, X. Correig, B. Bouchikhi, "Application of a portable electronic nose system to access the freshness of Moroccan sardines", Material Science and Engineering C, 28 (2008), pp. 666-670.

[19] Y. Blixt and E. Borch, "Using an electronic nose for determining the spoilage of vacuum-packaged beef". International Journal of Food Micro-biology, 46(2) (1999), pp. 123-134.

[20] S. Oshita, K. Shima, T. Haruta, Y. Seo, Y. Kawagoe, S. Nakayama, H. Takahara, "Discrimination of odors emanating from 'La France' pear by semi-conducting polymer sensors", Computers and Electronics in Agriculture 26 (2000), pp. 209-216.

[21] K-T. Tang, S.-W. Chiu, C.-H. Pan, H.-Y. Hsieh, Y.-S. Liang and S.-C. Liu, "Development of a portable electronic nose system for the detection and classification of fruity odors", Sensors 2010, 10, 9179-9193.

[22] E. Hines, E. Llobet, J. W. Gardner, "Neural network based electronic nose for apple ripeness determination", Electronic Letters, 35 (1999), pp. 821–823.

[23] A. Legin, A. Rudnitskaya, Yu. Vlasov, C. di Natale, E. Mazzone, A. D'Amico, "Application of electronic tongue for qualitative and quantitative analysis of complex liquid media", Sensors and Actuators B, 65 (2000), pp. 232–234.

[24] J. W. Gardner, H. V. Shurmer and T. T. Tan, "Application of an electronic nose to the discrimination of coffees", Sensors and Actuators B, 6 (1992), pp. 71–75.

[25] M. Ghasemi-Vrnamkhasti, S.S. Mohtasebi, M.L. Rodrigez-Mendez, J. Lozano, S.H. Razavi and H. Ahmadi, "Potential Applications of electronic nose technology in brewery", Trends in Food Science and Technology, 22 (2011), pp. 165-174.

[26] T. B Borjesson, T. Eklov, A. Jonsson, H. Sundgren and J. Schnurer, "Electronic nose for odor classification of grains", Cereal Chemistry, 73, 1996, pp. 457-461.

[27] R. Mariaca and J.O. Bosset, "Instrumental Analysis of volatile (flavour) compounds in milk and dairy products", Lait, 77 (1997), pp. 13-40.

[28] W.J. Harper, S. Sohn and K. Da Jou, "The role of fatty acids in the aroma profiles of swiss cheese as determined by an electronic nose", 3rd International Symposium on Olfaction and Electronic Nose, Toulouse, 1996.

[29] S. Breheret, T. Talou, B. Bourrounet and A. Gaset, "On-line differentiation of mushrooms aromas by combined Headspace/multiodor gas devices". In: P. Etievant and P. Schreier (Eds), Proceedings of Bioflavour 95. Dijon: INRA Editions, 1995, pp. 103-107.

[30] R. Splivallo, N. Valdez, N. Kirchoff, M. Castiella Ona, J-P Schmidt, I. Feussner and P. Karlovsky, "Intraspecific genotypic variability determines concentrations of key truffle volatiles", New Fytologist, vol. 194 (3) (2012), pp. 823-835.

[31] Barry Lee, "Taking Stock of the Australian Truffle Industry", RIRDC Publication No 08/124, RIRDC Project No PRJ-002643, July 2008.

[32] T. Talou, M. Delmas, A. Gaset, "Principal constituents of black truffle (Tuber melanosporum) aroma", *J. Agric. Food Chemistry*, 35 (1987), p. 774-777.

[33] P. Diaz, E. Ibanez, F.J. Senorans, G. Reglero, "Truffle aroma characterization by headspace solid-phase microextraction", Journal of Chromatography A, 1017 (2003), pp. 207-214.

[34] R. Spivallo, S. Ottonello, A. Mello, P. Karlovsky, "Truffle volatiles: from chemical ecology to aroma biosynthesis", New Phytologist, 189 (2011), pp. 688-699.

[35] T. Talou, A. Gaset, M. Delmas, M. Kulifaj, C. Montant, "Dimethyl sulphide: The secret for black truffle hunting by animals?", Mycol. Res. vol. 94 (1990), pp. 277-278.

[36] R.E. March, D.S. Richards, R.W. Ryan, "Volatile compounds from six species of truffle – head space analysis and vapor analysis at high mass resolution", Int. J. Mass Spectrom., vol. 249-250 (2006), pp. 60-67.

[37] R. Spivallo, M. Novero, C. Bertea, S. Bossi, P. Bonfante, "Truffle volatiles inhibit growth and induce an oxidative burst in Arabidopsis thaliana", New Phytologist, 175 (2007), pp. 417-424.

[38] R. Claus, H. O. Hoppen and H. Karg, "The secret of truffles; a steroidal pheromone?", Experientia, vol. 37, (1981) pp. 1178-1179.

[39] R. Splivallo, S. Bossi, M. Maffei, P. Bonfante, "Discrimination of truffle fruiting body versus mycelial aromas by stir bar sorptive extraction", Phytochemistry, vol. 68 (2007), pp. 2584-2598.

[40] V. Christopoulos, P. Psoma and S. Diamandis, "Site characteristics of Tuber magnatum in Greece", Acta Mycologica, vol. 48 (1) (2013), pp. 27-32.

[41] A. K. Jain, R. Duin, and J. Mao, "Statistical Pattern recognition: A Review", IEEE Transactions on Pattern Analysis and Machine Intelligence, Vol. 22 (1) (2000), pp. 4-37.

[42] J. Shlens, "A Tutorial on Principal Components Analysis", http://www.brainmapping.org/NITP/PNA/Readings/pca.pdf (last accessed: December 2013).

Biographies

Despina Zampioglou received the degree of IT & Communications Engineering at the Technological and Educational Institute of Central Macedonia, Serres, Greece, in 2013. She participated in the contest "Imagine Cup" of Microsoft with the electronic-nose project presented in this paper and received the 3rd prize in the category "Innovation". She is currently a fiber optics Engineer in Raycap Corporation.

John A. Kalomiros received the degree of Physics and a MS degree in Electronics from the Aristotle University of Thessaloniki, Greece. His PhD thesis is on the development of vision systems for robotic applications. His research interests include micro-electronic devices, microcontrollers, design of digital systems and machine vision. He is also working on systems for digital measurements and instrumentation. He is a faculty member in the Department of Informatics Engineering at the Technological and Educational Institute of Central Macedonia, Greece, where he teaches electronics and related subjects.

5

Data acquisition for ultrasonic transducer evaluation under spread spectrum excitation[*]

L. Svilainis, A. Chaziachmetovas, D. Kybartas and V. Eidukynas

Signal Processing Department, Kaunas University of Technology, Studentu str. 50-340, LT-51368, Kaunas, Lithuania, e-mail: linas.svilainis@ktu.lt

Abstract

This chapter is presenting the ultrasonic data acquisition principles and techniques used for the ultrasonic transducer properties investigation under spread spectrum excitation. Spread spectrum (SS) signals differ from conventional pulse, step and toneburst signals in that sense that they have spectral components spread in time. Thanks to this property signal can be compressed gaining both energy and resolution. Reasoning for the SS signals based on time of flight estimation theory is given. Discussion on signal and transducer types is given and governing equations are presented. Ultrasonic transducer as energy converter is presented and essential parameters outlined. Discussion is separated into two threads: i) electrical energy transduction into pressure investigation; and ii) directivity of the pressure field investigation. It was aimed to investigate the transducer transduction and directivity properties relation to excitation signal type. Spread spectrum signals and conventional pulse, step and toneburst signals were used in comparison. Setup and data acquisition techniques for transducer impedance measurement, interaction with excitation generator, power delivery and bandwidth optimization are discussed. Directions, system structure and setup for transducer directivity evaluation are given.

[*] This research was funded by a Grant (No. MIP-058/2012) from the Research Council of Lithuania.

V. Haasz and K. Madani (Eds.), Advanced Data Acquisition and Intelligent Data Processing, 83–126.

Keywords: signal transduction; impedance measurement; data acquisition system; ultrasonic transducer directivity; spread spectrum imaging; transducer parameters measurement.

5.1 Introduction

Ultrasound is addressed as acoustic or mechanical waves oscillating at frequencies above the human audible range. It can propagate in gases, liquids and solid materials, interacting with media particles. Thanks to the direct mechanical interaction it is able to extract the material properties which later can be used in measurement or imaging. Ultrasound is used in non-destructive testing (NDT) of materials for coating or layers thickness measurement; non-contact vibrations measurement; mechanical properties, load and density evaluation; chemical composition evaluation; non-destructive evaluation (NDE) [1–5]; non-contact temperature estimation [6]. Ultrasound is exploited for navigation and robotic vision in air and water [7, 8]; surface profiling; gas or liquid flow velocity measurement [9, 10]. It is used in biology for diagnostics [11, 12]. Food industry explores ultrasound for products quality management. Ultrasonic techniques offer simple and portable testing equipment which is safe and easy in exploitation [13, 14]. Ultrasonic transducer is the device used to convert the electrical energy into pressure signal and vice versa [15].

Our focus is concentrated on ultrasound applications in measurement and imaging. Two major tasks can be outlined here: i) imaging or detection of discontinuity in the material under investigation and rectification of its coordinates and reflection strength ii) measurement of propagation time and amplitude of the signal to obtain the velocity, attenuation or material thickness. Such measurements can be used to derive the related mechanical properties of the test object or create the corresponding map which later is related to its status. Both coordinates location and propagation time estimation can be adjoined to the same problem: estimation of the delay of the signal traveled in test material. This time delay is addressed as time of flight (ToF), time of arrival (TOA), time delay estimate (TDE) or time delay of arrival (TDOA). Accuracy of ToF estimation is defined by signal parameters: energy, noise and frequency band [16]. Ultrasonic transducer in conjunction with electronics limits these parameters [17]. Signal is also altered by the material attenuation and scattering [1]. Measurement and especially imaging applications also require signals separation both in temporal and spatial domain which is usually addressed as resolution [18,

19]. Temporal resolution is defined by signal duration or envelope bandwidth: the wider is the signal spectrum the narrower is the signal hence the resolution is better. Spatial resolution is defined by ultrasonic transducer directivity: beam divergence angle defines the resolution.

Raising the energy of the signal and the bandwidth of are contradicting: in case of high energy signals [20, 21] signal duration is long so resolution is reduced. To solve this contradiction the spread spectrum (SS) signals can be used which are compressible [22, 23, 24].

Both energy and bandwidth are important for SS signals [10, 25]. Therefore investigation of transducer performance under such complex excitation signal is needed [26, 27]: transducer must match the bandwidth of the excitation signal [28, 29]. In order to evaluate these parameters specific data has to be collected and processed. Signal and data acquisition and processing aspects related to transducer and electronics are discussed below. Transducer impedance interaction with excitation circuitry and attainable efficiency and bandwidth are evaluated. Directivity measurement procedures with the spectral content changes investigation are presented.

5.2 Spread spectrum signals

Conventional ultrasonic systems are using pulse signals. Reasoning for such choice is that pulse signals exhibit wide spectrum, generation of such signals is simple so the size of equipment is small [30]. Another important property is that such signals possess low correlation sidelobes and occupy short duration which contributes the temporal resolution. But pulse signals have several essential drawbacks [31]: i) low energy; ii) limited spectral content control; iii) presence of spectral zeros. Spectral content can be varied only by adjusting the duration. Duration must be short when aiming the bandwidth but in such case signal energy is reduced [20]. Energy can be improved by increasing the amplitude of the pulse but there are certain limits in transducer and electronics capabilities.

Estimation of the ToF is the most recently used procedure in various ultrasonic applications. The received signal $s(t)$ can is treated as delayed by *ToF* and attenuated version of the reference (transmitted) signal *ref(t)* with white noise component $n(t)$ added. Noise $n(t)$ can be assumed a non-correlated additive white Gaussian noise with power spectral density N_0. The goal of the ToF measurement is to find an estimate of the *ToF* from this corrupted signal. Advanced techniques use maximum likelihood criteria for time position estimation [32–34]. Likeness between the reference (transmitted) signal and received signal can be evaluated using the cross-correlation maximum, minimum of difference L1 norm or minimum of difference L2 norm [25]. Reference signal can be pre-recorded from some reference object or produced using adaptive model [35].

Cross-correlation technique is using the peak position of the cross-correlation function (CCF) x as the ToF estimate [36]:

$$ToF_{CC} = \arg[\max(x(\tau))], \quad x(\tau) = \int_{-\infty}^{\infty} s_R(t) \cdot s_T(t-\tau) dt . \quad (5.1)$$

The difference remainder $L1$-norm minimization technique is using the minimum of magnitude of the reference s_R and received s_T signals [36]:

$$ToF_{L1} = \arg\{\min[L1(\tau)]\}, \quad L1(\tau) = \int_{-\infty}^{\infty} |s_R(t) - s_T(t-\tau)| dt . \quad (5.2)$$

The $L2$-norm minimization technique is using the delay τ where $L2$-norm (power) of s_R and s_T difference is minimal as the ToF estimate:

$$ToF_{L2} = \arg\{\min[L2(\tau)]\}, \quad L2(\tau) = \int_{-\infty}^{\infty} [s_R(t) - s_T(t-\tau)]^2 dt . \quad (5.3)$$

The difference is whether peak or minimum value is used (Figure 5.1).

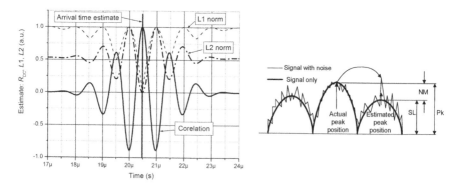

Figure 5.1 Time of flight estimation techniques (left) and sidelobes influence (right)

If estimation is done by locating the correlation function's peak [25] in the presence of noise with power spectral density N_0, the variance of ToF is defined by Cramer-Rao lower error bound [16, 37]:

$$\sigma(TOF) \geq \frac{1}{2\pi F_e \sqrt{SNR}}, \quad SNR = \frac{2E}{N_0}, \quad (5.4)$$

where E is signal $s_I(t)$ energy, F_e is effective bandwidth of the signal and noise density is estimated from electronics model or measured [38]:

$$E = \frac{1}{Z_{ADC}} \int_{-\infty}^{\infty} |s(t)|^2 dt = \frac{2}{Z_{ADC}} \int_{0}^{\infty} |S(f)|^2 df, \quad N_0 = \frac{e_{ntot}^2}{Z_{ADC}}. \quad (5.5)$$

The effective bandwidth F_e is a square sum of envelope bandwidth β and center frequency f_0:

$$F_e^2 = f_0^2 + \beta^2 = \frac{\left[\int_{-\infty}^{\infty} f |S(f)|^2 df \right]^2}{E^2} + \frac{\int_{-\infty}^{\infty} (f - f_0)^2 |S(f)|^2 df}{E}. \quad (5.6)$$

Increasing popularity of composite materials demands for NDT of the composite goods. Ultrasonic signal suffer attenuation and scattering in composite. Split spectrum imaging [39, 40], exploits frequency domain diversity but demand wide bandwidth. Signal transmission losses are high in the air-coupled ultrasound applications [41]. Equations (5.4)-(5.6) indicate that accuracy is improved by increasing the signal bandwidth or increasing the energy. Energy enhancement is possible by a probing signal amplitude or duration increase. The excitation amplitude is limited by the transducer construction and the electronics capabilities. Another approach would be to increase the duration of the exciting signal. But if it is a simple CW burst then the envelope of the correlation function is not sharp and resolution will be low. Imaging systems demand high envelope bandwidth to ensure narrow correlation mainlobe. Low sidelobes are important to avoid the artifacts on the image. Furthermore, high level of sidelobes can produce the abrupt errors of ToF estimation (Figure 5.1, right). Therefore not only peak value (Pk in Figure 5.1, right) but also the level of sidelobes (SL) and resulting noise margin (NM) are important [20]. If application requires wide bandwidth and high signal-to-noise ratio (SNR) then use of SS signals is attractive: these signals can offer both the bandwidth and the energy [29, 31, 42] and can be compressed using matched filter [22, 23]. Compression is defined by duration-bandwidth product [19]:

$$\tau_e \cdot \beta = \frac{\int\limits_{-\infty}^{\infty}(t-t_0)^2|S(t)|^2\,dt}{E} \cdot \frac{\int\limits_{-\infty}^{\infty}(f-f_0)^2|S(f)|^2\,df}{E}. \qquad (5.7)$$

Variety of techniques exists that produce the SS signal: arbitrary waveform, phase manipulated sequences, linear and non-linear frequency modulated signal (chirp) and their binary counterparts (Figure 5.2).

Arbitrary waveform signals offer most flexibility but are not easy to generate [29, 31]. Chirp or linear frequency modulation signals are straight forward implementation of spread spectrum idea: their frequency components are distributed over time. Note the spectral differences of pulse continuous wave (CW) toneburst and chirp spectrograms (Figure 5.3).

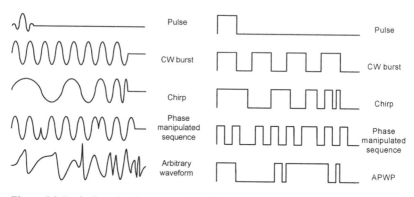

Figure 5.2 Excitation signal types: analog (left) and square wave counterparts (right)

But the correlation sidelobes (Figure 5.4) of the chirp are higher than for single rectangular pulse [31]. Nonlinear frequency modulation signals are gaining popularity thanks to the ability to control the shape of the spectrum and so the sidelobe level [29]. In turn, rectangular pulses are easy to generate and exhibit low correlation sidelobes, but do not possess the high energy and coding possibility as SS. Novel spectrum spread technique was suggested in [31, 43, 44]: trains of the arbitrary position and width pulses (APWP, Figure 5.2, right). Once APWP signal is constructed by the set of constant amplitude but arbitrary duration pulses with arbitrary spacing, generation of such signals should be as simple as for rectangular pulses. Initial study carried out in [31] suggests that APWP can have lower than rectangular pulse correlation sidelobes. APWP signal energy is the same as chirp signal which correlation sidelobes are much higher.

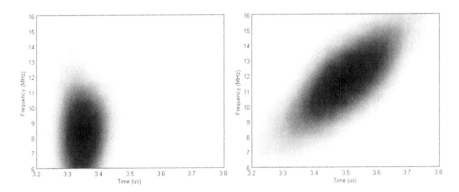

Figure 5.3 Spectrograms of example signals: pulse (left) and chirp (right)

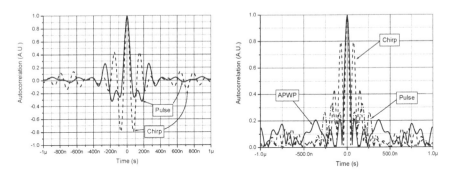

Figure 5.4 Comparison of chirp and rectangular pulse sidelobes in RF correlation function (left) and correlation function envelope with correlation function for APWP added (right)

SS signals require wide bandwidth of the transmission channel. But essential part of the ultrasonic system, the transducer, limits the SS performance: transducer must match the bandwidth of the excitation signal [28, 29]. Therefore investigation of transducer performance under complex excitation signal is needed [45, 46] to evaluate transducer interaction with electronics for attainable efficiency and bandwidth.

5.3 Ultrasound transduction performance

Main function of the ultrasonic transducers is to convert the power from electrical port into mechanical energy and vice versa. In most cases transducers are reciprocal devices and can be used for transmitting pulses into media and receiving scattered echoes.

Electrical and mechanical losses in transducer can be described by electrical (Q_e) and mechanical (Q_m) quality factors. Therefore overall performance of transducer used for energy conversion requires high electromechanical coupling and low losses.

Transduction efficiency can be linked to electromechanical coupling factor, which relates the energy converted from electrical port into mechanical energy $W_{El\text{-}Mech}$ to total electrical energy supplied W_{El} when in transmission mode and energy converted from mechanical into electrical $W_{Mech\text{-}El}$ to total mechanical energy W_{Mech} when in reception mode:

$$k = \sqrt{\frac{W_{El \rightarrow Mech}}{W_{El}}} = \sqrt{\frac{W_{Mech \rightarrow El}}{W_{Mech}}} . \qquad (5.8)$$

This factor depends on mechanical and electrical properties of the transducer material [47]. Consequently, ultrasonic transducer frequency, bandwidth and efficiency depend on it. The most common materials use the piezoelectric effect. Several materials can be used here: quartz, barium titanate ($BaTiO_3$), lead metaniobate ($PbNb_2O_3$), lead zirconate titanate (PZT) and Piezoelectric polymers (e.g. polyvinylidene fluoride, PVDF) [15, 48]. Microelectromechanical systems (MEMS) technology can also be used to ultrasonic sensors: capacitive micromachined ultrasonic transducers (CMUTs) offer wide bandwidth, automated manufacturing and integration with electronics [49]. Magnetostrictive material can be used: it changes its length under magnetic field; and magnetic field is created by mechanical stress. Such transducers are efficient (high Q_m) but can operate only up to hundreds of kilohertz and in narrow frequency range which is strongly dependent of length of magnetostrictive rod. Therefore a use of magnetostrictive transducers with wideband signals is limited.

In case of non-contact ultrasonic inspection waves into material are injected without immersion or direct mechanical attachment using optical, electromagnetic and air-coupled techniques. Optical excitation is using thermoelastic regime or ablation. Optical reception can use interferometry, optical beam deflection or knife edge modulation. Electromagnetic Acoustic Transducer (EMAT) is actuated electromagnetically [50]. EMAT operation is based on induction of eddy current in conductive specimen by magnetic coil. Another part of EMAT is permanent magnet. Lorenz force creates mechanical stress in a region with induced currents and acoustic wave is transmitted. EMAT transducers are broadband but can only be used with conductive materials. Disadvantage is low EMAT efficiency: received signal is few microvolts and excitation currents are hundreds of amperes.

5.3.1 Transducer frequency response

Efficient transmission and reception of spread spectrum signals which usually are wideband is possible only when bandwidth Δf of transducer is relatively wide compared to its center frequency f_0 ($\Delta f/f_0 > 50\%$). Aiming the economy and bandwidth, piezoelectric transducers are the best choice when spread spectrum excitation is targeted [51]. Piezoelectric transducers have simple construction, are small, cheap and usually offer high sensitivity.

Typical piezoelectric transducer contains a piezoelement which has two electrodes attached (Figure 5.5).

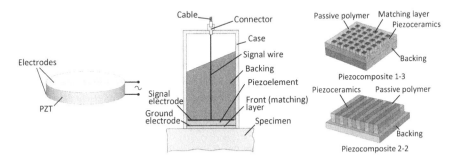

Figure 5.5 Piezoelectric transducer: piezoelement (left), construction (center) and piezocomposite elements (right)

Dimensions of piezoelement are changing depending on the voltage applied on electrodes. Typical displacements of piezoelement are in range

10^{-10} m/V. Ceramic-based piezomaterials have mechanical and electrical quality factors Q_m, Q_e > 30-50. Thus, frequency range of efficient response for single element is narrow.

Excitation of acoustical waves in wide frequency range is possible using three ways: i) wideband impedance matching layers [52], ii) use of composite piezoceramic elements [53, 54], iii) electrical matching [55]. All three approaches are used simultaneously in transducers design.

Transmission of ultrasonic pulses into specimen is through the front layer (Figure 5.5, center). This layer serves for protection and for acoustical impedance matching in order to avoid pulses scattering from border with different acoustical impedance.

Acoustical impedance of the material relates the applied stress σ and particle velocity v_m in media:

$$\sigma = Z_a \cdot v_m . \tag{5.9}$$

It depends on density of material ρ and on speed of sound c:

$$Z_a = \rho \cdot c . \tag{5.10}$$

Most of piezoelectric transducers are based on piezoceramics, such as PZT [51]. Typical acoustic impedance of such ceramics is $Z_{PZ} \approx 30$-40 MRayls. Some materials have lower acoustical impedances (water $Z_a \approx 1.5$ MRayl, aluminum $Z_a \approx 17.5$ MRayl) so an acoustical matching of impedances is necessary. There is no need for impedance matching when transducers are used for steel materials ($Z_a \approx 40$-50 MRayl).

Typical acoustical matching is based on transmission lines theory. A matching layer is used in front of piezoelement (Figure 5.5). Most simple applications use single matching layer. Thickness of this layer is selected close to $\lambda_0/4$, where wavelength is defined by frequency used f_0 and speed of sound in matching layer c_{ML}:

$$\lambda_0 = c_{ML} / f_0 . \tag{5.11}$$

To match the piezoelement impedance Z_{PZ} to the material Z_a the acoustical impedance of the matching material is selected equal:

$$Z_{ML} = \sqrt{Z_a \cdot Z_{PZ}} . \tag{5.12}$$

It may appear, that matching layer will have a narrow band but such system of resonant tanks will be influenced by coupling and can be tuned by varying the layer thickness and acoustical properties (Figure 5.6).

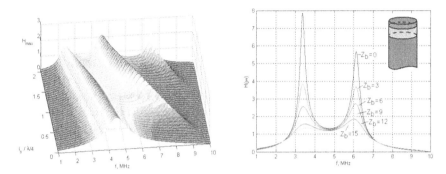

Figure 5.6 Transfer function of the transducer: influence of matching layer thickness (left) backing material acoustical impedance influence (right)

Backing layer serves as mechanical damping of piezoelement from back side and allows increasing the amplitude of the acoustical wave radiated into specimen. Varying the acoustic impedance of the backing (Figure 5.6, right) it is possible to get a flat transfer function at the expense of reduced efficiency. Increasing the number of matching layers can further increase the bandwidth and improve the sensitivity. Some approaches allow using of slightly different matching layer thickness [56, 57]. Transducer with two matching layers gives more degrees of freedom for optimization, such as different thicknesses and impedances of the matching layers [15].

Designing piezoelectric transducers to meet particular performance requirements is a complicated and time consuming engineering process. Parameters such as power, sensitivity and bandwidth depend on highly complex mechanical and piezoelectric material properties, part size and shape, and other electrical and mechanical parameters. Difficulties are compounded when transducers must operate in wide frequency range.

A simple way for such transducers design is to use a one-dimensional (1D) equivalent circuit models. These 1D tools provide approximations of transducer behavior based on simplified, lumped circuit representations of transducers using inductors, capacitors and resistors [58–60]. The resulting models approximately represent the true behavior of transducer at single mode of operation. The limitation of 1D model is the existence of other vibrational modes and harmonics in real piezoelements and matching layers. There are lateral or radial modes in piezoelement when part of energy is transformed into transversal displacements. Such energy circulation in transducer limits change rate of excitation frequency.

Using the direct coupled-field analysis capabilities of Finite Elements Analysis (FEA) software, engineers have implemented a better approach based on finite element analysis to quickly and effectively arrive at optimal transducer designs without the delays, guesswork and inaccuracies of other methods [56, 61]. FEA utilizes a full 3D simulation of the transducer with piezoelectric, mechanical and acoustic parts to characterize the dynamic responses of the transducer. FEA is offering better possibilities of optimization of frequency response. Fluid structure interaction (FSI) and acoustic elements model water-loaded behavior in determining attributes such as frequency-dependent beam patterns, directivity, transmit power and receive sensitivity [62].

Matching layer can be diced in order to avoid the frequency response distortions caused by lateral modes [62]. Another approach [63] is to dice both the piezoelement and the matching layer. Gaps are filled by passive polymer, reducing the crosstalk between the dices. Such arrangement is addressed as composite transducer. In general any type of combination of piezoceramics with non-piezoelectric material is the piezocomposite. Possible combinations usually are addressed by two digit number where first digit is used to express the amount of axes along which composite is coupled while second digit express the same for passive polymer. For instance structure 1-3 means that composite is linked along one axis while passive material is linked along all three axes (Figure 5.5, right). Another popular arrangement is 2-2 (Figure 5.5, right). It is possible to achieve the summation of thickness and lateral modes oscillations by properly selecting the dicing dimensions [64]. Varying the volume fraction, shape of inclusions different effective properties are obtained. One of the essential results is the reduction of acoustical impedance. This results in electromechanical coupling increase which in turn improves the transducer efficiency. Another improvement is the electrical impedance reduction which allows using lower excitation voltages. Essential for SS is that these modifications significantly increase the transducer bandwidth.

Impedance of piezoceramic element at single vibrational mode can be described by Butterworth-Van-Dyke (BVD) equivalent circuit (Figure 5.7, left). The electrical capacitance of the transducer is represented by C_0. The mechanical boundary conditions are modeled by R_m, and C_m, while the mass of the mechanical system is described by inductance L_m. In general frequency response of piezoelement is similar to capacitive element impedance (Figure 5.7, right) but it has serial (f_r) and parallel (f_a) resonance frequencies. Such approximation is correct only for first mechanical resonance omitting higher harmonics and vibration modes. More accurate electrical models are based on transmission line theory (Mason, KLM and other, [58, 59, 65, 66]).

An efficiency of piezoelectric energy conversion is expressed using dynamic electromechanical coupling factor [47]:

$$k_{eff}^2 = \frac{f_a^2 - f_r^2}{f_a^2} . \tag{5.13}$$

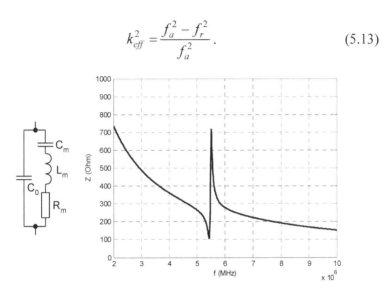

Figure 5.7 Van-Dyke model of piezoelement (left) and its frequency response (right)

The most efficient transmission is on serial resonance frequency f_s which correspond to mechanical resonance frequency. This frequency can be changed using electrical and mechanical matching [56]. Frequency of efficient transmission can be shifted down or broadened using mechanical loading of piezoelement or by use of acoustical matching layers. Compensation of parallel capacitance C_0 allows efficient excitation in wide frequency range. Therefore, properly designed piezoelectric transducers can transmit and receive wideband ultrasonic signals.

5.3.2 Data acquisition for transducer impedance measurement

Transducer impedance Z_T is an important parameter [67, 68] characterizing it as an electronic circuit and is represented as a complex quantity versus frequency. Variety of techniques for impedance estimation exists: transmission/reflection, I-V, RF I-V and auto-balancing bridge (ABB) [69]. Basing the usual range of ultrasonic transducers (20 kHz to 50 MHz) and desired uncertainty of 10 % two candidate techniques remain: I-V and the ABB. For frequencies above 1 MHz the RF I-V technique is suitable [70]. The I-V technique allows for 100 Hz to 100 MHz frequency range coverage and can be easily arranged using plain high impedance amplifier (Figure 5.8) and general purpose data acquisition system. Advantage of such setup is not only its simplicity but also availability of ground connection: transducer connector is usually asymmetric type.

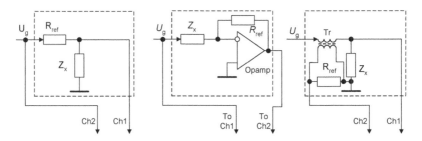

Figure 5.8 Measurement setup for I-V (left), ABB (center) and RF I-V (right) techniques

The ABB (Figure 5.8, center) employs the operational amplifier inverting topology [71]. Virtual ground is produced on inverting input so voltage drop on reference impedance is not supplied to the measurement point. This technique offer wider probing frequency and values range. But it has the disadvantage: only virtual ground is available.

In both cases acquisition system must contain the excitation signal generator which frequency is derived from the sampling frequency of analog-to-digit converters (ADC). In such case the amplitude and phase extraction form the measured signals can be simplified since frequency of the excitation signal and the sampling rate of the acquisition are fixed. If probing signal contains only single frequency f then sine wave correlation (SWC) technique [72] can be applied to extract the signal amplitude and phase. Fitting the function (5.14) to the signal $y_1 \ldots y_M$, acquired at a frequency f_s at time instances $t_1 \ldots t_M$ then is accomplished as:

$$U_c = \frac{\sum_{m=1}^{M}[\cos(2\pi f t_m)\cdot y_m]}{\sum_{m=1}^{M}[\cos(2\pi f t_m)]^2}, U_s = \frac{\sum_{m=1}^{M}[\sin(2\pi f t_m)\cdot y_m]}{\sum_{m=1}^{M}[\sin(2\pi f t_m)]^2}. \quad (5.14)$$

Then the measured signal magnitude and phase is:

$$U = \sqrt{U_c^2 + U_s^2}, \varphi = \arctan\left(\frac{U_s}{U_c}\right). \quad (5.15)$$

Investigation [72] suggests that both techniques can be used sparingly for 100 kHz to 20 MHz frequency range. If ultrasonic transducer cable is long, then impedance estimation will influence by test fixture errors [73]. The imperfectness of the measurement circuit can be compensated by [69]:
- Open/Short measurement;
- Open/Short/Load measurement;
- conversion via cable parameters.

Compensation using additional Open (Z_o), Short (Z_s), Load (Z_{stdm}), measurement improves results significantly [73]:

$$Z_{DUT} = Z_{std}\frac{(Z_o - Z_{stdm})\cdot(Z_{xm} - Z_s)}{(Z_{stdm} - Z_s)\cdot(Z_o - Z_{xm})}, \quad (5.16)$$

where Z_o, Z_s, Z_{stdm} are measured impedances obtained by open, short and known impedance Z_{std} conditions accordingly, Z_{xm} measured impedance.

Amplitude of the probing signal alters the measured impedance [74]. Then, it is desirable to measure the impedance under planned excitation conditions (high voltage). It is also desirable to perform the measurement fast. Pulsed impedance measurement [75, 76] solve both problems. Instead of probing the transducer with low voltage harmonic waves, short high voltage pulse can be used. Only I-V or RF I-V setup can used since high voltage could damage the amplifier used in ABB circuit. Setup has to be modified adding voltage dividers. RF I-V setup has an advantage: current measurement channel is isolated and only one divider is needed.

Pulsed measurement has one disadvantage: signal-to-noise ratio is low since signal duration is short.

Application of spread spectrum signal should improve the situation [77]. Experimental investigation has been carried out to compare the RF

I-V setup performance under harmonic excitation, pulse and chirp (SS) excitation. Impedance of the wideband composite 5 MHz transducer C543 from Olympus was measured using impedance analyzer 65120B from Wayne Kerr using low voltage harmonic excitation and experimental RF I-V setup using high voltage excitation from ultrasonic pulser [42]. No Open/Short/Load compensation was used in order to have the ability to compare techniques performance on raw data. It can be seen that results for all signal types match closely (Figure 5.9).

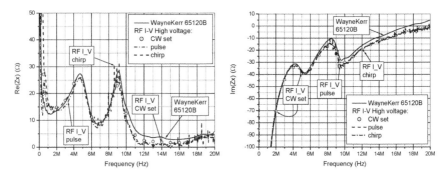

Figure 5.9 Comparison of RF I-V measurement results for CW, pulsed and chirp (SS) probing signals: real (left) and imaginary (right) part of 5MHz transducer C543 impedance

There is a slight variation in the impedance obtained for chirp signal with periodicity of 0.33 MHz. This could relate to rectangular excitation signal or its duration. Conclusion can be drawn that RF I-V topology is a simple and convenient technique for pulsed signal impedance estimation. Both pulse and spread spectrum signals can be applied.

5.3.3   Matching the excitation generator and transducer

In general, power delivery to load is maximized when load impedance is real and equal to generator intrinsic resistance R_g. It is evident that if transducer impedance is complex the power supplied to transducer will be complicated and excitation efficiency is reduced [67, 78].

If imaginary parts L_m, C_m of transducer BVD model are in resonance the equivalent transducer circuit is simplified to parallel connection of R_m and C_0 (Figure 5.10). If compensation is needed just for narrow frequency range C_0 can be compensated by parallel or serial inductance [55].

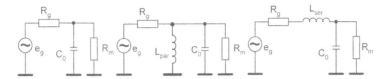

Figure 5.10 Equivalent transducer circuit in (left to right): case of the mechanical resonance; suggested matching inductance connection in parallel and series

Parallel L_{par} inductance value for desired angular frequency ω_s is:

$$L_{par} = -\frac{X_{par}\big|_{\omega=\omega_s}}{\omega_s}, \tag{5.17}$$

where R_{par} and X_{par} are obtained from serial transducer impedance model:

$$X_{par} = \frac{R_T^2}{X_T}\left[1+\left(\frac{X_T}{R_T}\right)^2\right], \, R_{par} = R_T\left[1+\left(\frac{X_T}{R_T}\right)^2\right]. \tag{5.18}$$

Inductor quality will affect the compensation results. If high inductor quality Q is achieved ($\sim$100) then equivalent inductance L_{xp} :

$$L_{xp} = L_{xs}\left(1+\frac{1}{Q_L^2}\right), \; r_{plossc} = r_{sloss}\left(1+Q_L^2\right), \tag{5.19}$$

is close to its series equivalent L_{xs} and thanks to parallel resonance C_0 and L_{xp} are resonated out and only parallel connection of inductor loss resistance r_{ploss} and transducer emission resistance R_m remain [79].

For serial inductance compensation X_T of transducer impedance is used

$$L_{ser} = \frac{X_T}{\omega_s}. \tag{5.20}$$

Again, inductor quality represented by serial inductor losses resistance r_{sloss} has to be taken into account as addition to generator's R_g.

Use of "*L*" [80] matching circuit is implemented by placing two reactive components between amplifier and the transducer (Figure 5.11).

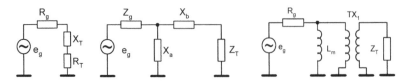

Figure 5.11 Impedance (left) matching by "L" circuit (center) and transformer (right)

The corresponding reactance of matching circuit is:

$$X_a = -\frac{R_g^2 + X_g^2}{QR_g + X_g}, X_b = QR_T - X_T,$$ (5.21)

where X_g is reactance of generator and Q is:

$$Q = \pm\sqrt{\frac{R_g}{Re(Z_T)}\left[1+\left(\frac{X_g}{R_g}\right)^2\right]-1}.$$ (5.22)

Taking both Q solutions and swapping source and load positions four configurations are available. Generator impedance can be complex.

Application of *transformer matching* (Figure 5.11) is a popular technology in RF technique. Idealized transformer presented in Figure 5.11 is characterized only by turns ratio *n* and the magnetizing inductance L_m. Such circuit is not complete. For extensive analysis full transformer model should be used [81]. To maximize the power delivered to transducer the magnetizing inductance L_m should be chosen ten times larger the source impedance and the transformer turn's ratio are:

$$L_m = \frac{10R_g}{\omega_s}, n = \sqrt{\frac{|Z_T|}{R_g}}.$$ (5.23)

Transformer magnetising inductance L_m can also be used to match the capacitive part of the transducer impedance using (5.17).

Results of the above described techniques for air-coupled ultrasonic transducer impedance matching [79] are presented in Figure 5.12. Original impedance curves are labeled as "Orig", serial and parallel inductance matching results are labeled "Ser" and "Par" correspondingly, "L" label is for "L" matching and "Tr" for transformer matching. Solid arrow indicates the desired operation frequency where match is required.

Graphs in Figure 5.12 do not provide evident advantage of matching techniques. Several quantitative evaluation criteria were suggested in [38, 67, 82, 83]: power delivered to load at operation frequency; effective bandwidth; power delivery to load efficiency; power factor etc.

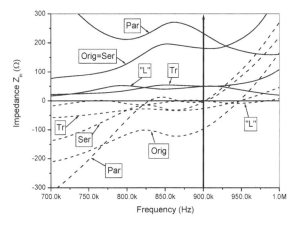

Figure 5.12 Matching circuit influence on transducer impedance

Complex power delivered to load can be calculated using the modified input impedance:

$$S_T = \frac{e_g^2 Z_{in}}{\left(R_g + Z_{in}\right)\left(R_g + Z_{in}^*\right)},$$ (5.24)

where Z_{in} is the input impedance after matching circuit was applied, Z_{in}^* is complex conjugate, e_g is generator open-circuit voltage.

Assuming that losses in matching circuit are negligible the real part P_T of this complex power S_T is equal to the power dissipated in real part of transducer acoustic impedance [84]. As it was mentioned before, this power (Figure 5.13) can be assigned to the acoustic emission of transducer [68].

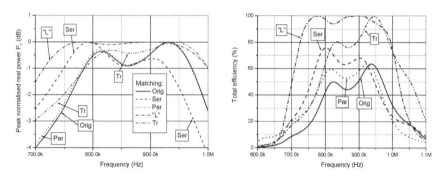

Figure 5.13 Normalized real power (left) and power delivery to load efficiency

Assuming that real part P_T of the complex power S_T is equal to the acoustical emission, power delivery efficiency can be estimated as:

$$\eta = \frac{4R_g \, \mathrm{Re}(S_T)}{e_g^2}100\% = \frac{4R_g Z_{in}}{(R_g + Z_{in})(R_g + Z_{in}^*)}100\% . \quad (5.25)$$

Refer to Figure 5.14 for source output impedance influence on power delivery efficiency for transducer IRY220.

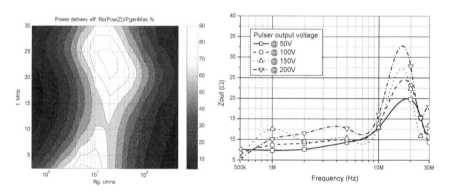

Figure 5.14 Influence of the pulser output impedance on power delivery efficiency for 20 MHz transducer IRY220 (left) and pulser output impedance AC response (right)

It can be seen that power delivery to load is maximized when pulser output impedance is around 20 Ω. With the measured transducer impedance available, transducer performance can be estimated. Impedance measurement allows to predict that moderate output impedance (20 Ω for investigated case) is needed in order to assure wideband SS signal match.

Once bandwidth and power delivery efficiency are defined by the high voltage pulser output impedance, investigation of pulser output impedance and attainable bandwidth has to be carried out. Standard EN_12668-1 [85] defines the ultrasonic pulser output impedance Z_o measurement procedure by using output voltages obtained at 50 Ω (V_{50}) and 75 Ω (V_{75}) pulser load:

$$Z_o = \frac{50 \cdot 75 \cdot (V_{75} - V_{50})}{75 \cdot V_{50} - 50 \cdot V_{75} N_0}.$$ (5.26)

Procedure can be slightly modified in order to investigate the output impedance at particular frequency. Data acquisition system used is capable of automated data collection using several excitation signals, then set of CW burst signals can be formed and stored for automated generation and acquisition. Voltages V_{50} and V_{75} then are recorded using 1:100 voltage divider formed by 5 kΩ and 50 Ω resistors. The amplitude and phase of the harmonic signal can be extracted by using the SWC technique. The CW bursts used were 10 periods long with frequency fixed in proportion to system clock frequency. Output impedance magnitude AC response for the pulser used in signals acquisition is presented in Figure 5.14 right. It can be seen that output impedance magnitude is stable with output voltage and is rising only at high frequency end. Variation is within 5 Ω to 21 Ω range. Obtained output impedance now can be used for pulser performance together with ultrasonic transducer estimation.

5.3.4   Transduction measurement

The frequency response of the transducer was determined by recording the probing signal on transducer clamps and the signal received from reflector. Taking the Fourier transforms of the two and dividing the received signal $u_{out}(f)$ by transmitted spectrum $u_{in}(f)$ transmission frequency response can be obtained:

$$G(f) = \frac{u_{out}(f)}{u_{in}(f)}.$$ (5.27)

Results for transmitted and received signals spectrum and resulting transmission obtained using (5.28) for focused wideband 10 MHz transducer IRY210 are presented in Figure 5.15. Transducer was excited by

SS signal (linear chirp, 1 μs duration, with frequencies ranging from 2.5 MHz to 20 MHz). Excitation signal amplitude was +/-30 V. It can be seen that transducer response follows the excitation bandwidth and SNR is high even beyond the bandwidth where excitation components are present. It can be seen that significant SNR is obtained for SS excitation so all the range used in excitation is usable even beyond the passband. Probing using pulse signal has a disadvantage of spectral zeros which produce false results due to division (Figure 5.15, right).

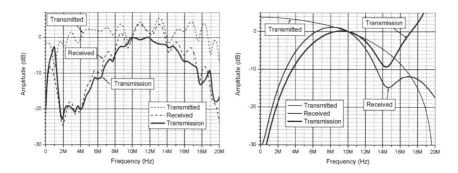

Figure 5.15 Transmitted, received signal spectrum and resulting transmission for 10 MHz transducer IRY210 when probed by SS (left) and pulse (right) signal

Acoustic pressure can be assessed using hydrophone. Spherically focused ultrasonic transducer (model TS 12PB2-7P30 from Karl Deutsch; (1-6) MHz frequency range, diameter 12 mm) was investigated using HNP-1000 hydrophone from Onda Corp., Sunnyvale. Hydrophone sensitivity M was corrected using the hydrophone C_h and input capacitance C_{in} [86]:

$$M_c(f) = \frac{M(f)C_h(f)}{C_h(f) + C_{in}} \quad . \tag{5.28}$$

Power delivery efficiency can be calculated from measured transducer input impedance for specific generator intrinsic impedance. Results for the same TS 12PB2-7P30 transducer are presented in Figure 5.16 for generator intrinsic impedance R_g=50 Ω and R_g=1 Ω.

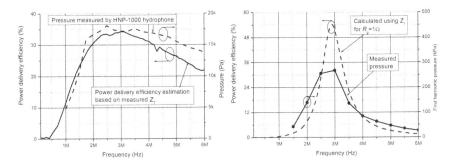

Figure 5.16 Transduction comparison: power delivery efficiency vs. pressure for R_g=50 Ω and U_g=10 V (left) and R_g=1 Ω and Ug=100 V (right) for CW toneburst excitation

Comparison of the results of impedance-based transduction estimation and measured pressure indicate that there for this type of transducer impedance measurement results can be used to predict the pressure AC response. Measured pressure peak is not as sharp as predicted by impedance measurement, because of distortion increase at higher pressures: some energy is transferred to higher frequencies. It can be concluded that this type of transducer cannot be excited by SS signals if generator output impedance is low: bandwidth is getting narrow. But it is suitable for 50 Ω output impedance pulser for SS excitation.

5.4 Transducer directivity performance

Transduction of the signal from electrical into mechanical port has to be accomplished by transducer directivity. It is the directivity that defines the spatial performance of the imaging system. Beam size at particular distance from transducer together with temporal response will define the attainable resolution of the images obtained [87]. This part is dedicated to transducer directivity evaluation.

5.4.1   Directivity estimation principles

An acoustical field of ultrasonic transducer is a result of superposition of pressure waves coming from all the points of radiating surface. Acoustic pressure p at specific distance r from transducer piston can be calculated using Rayleigh integral [1]:

$$p(r) = \frac{\omega\rho}{i \cdot 2\pi} \int_A v_M(x', y') \frac{e^{ikr'}}{r'} dx'dy', \qquad (5.29)$$

where ρ is density of acoustic specimen, ω – angular frequency, v_M – speed of motion of transducer aperture A point with coordinates x', y' and r' is distance from surface element dA to point where p is calculated (Figure 5.17, left). Phase of wave arrived from each surface element depends on wave number k and propagation distance r'. An amplitude of acoustical pulse depends also on distribution of mechanical displacement on radiating surface which can be expressed by $v_M(x',y')$.

Solution of this integral requires simplification of the task. A pressure distribution in specimen can be found only for circular aperture A and uniform distribution of displacement at surface. Analytical solution for pressure distribution for all distances is available only on axis of the beam (Figure 5.17, right). It can be seen that pressure varies near the transducer in so-called near field zone. These variations are caused by propagation path differences from several segments of transducer piston. Beyond the near field phase differences caused by unequal propagation paths are insignificant and on-axis pressure starts to decrease monotonically. This point defines the end of the near field and start of the far field zone. Ultrasonic field distribution in x and y directions is called beam profile.

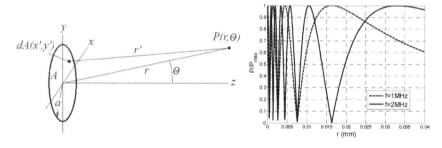

Figure 5.17 Graphical representation geometry used for calculation of acoustic field of transducer (left) and analytical solution result on axis for 10mm circular transducer (right)

Analytically it can be found only in far field zone. Half of beamwidth $\Delta\theta$ of the ultrasonic beam at first zero of main lobe and at half power is:

$$\Delta\theta_0 \approx \arcsin\frac{3.83}{ak}, \Delta\theta_{0,5} \approx \arcsin\frac{1.62}{ak}. \qquad (5.30)$$

If transducers aperture diameter is significantly larger than wavelength λ then beamwidth (in radians) can be calculated using simplified equation:

$$\Delta\theta_{0,5} \approx \frac{1.62}{ak} = 0.26\frac{\lambda}{a}. \qquad (5.31)$$

It is evident that beamwidth is directly related to the wavelength or signal frequency. Such approach is valid only for continuous wave (CW). Use of short pulse excitation differs from CW mode. Main difference is due the traveling time of acoustical pulses from the center and the edge of transducer. There is no analytical solution for this task and computational methods should be used to calculate interference of waves. Numerical integration methods are used to solve Rayleigh integral. They are based on analytical treatment of displacement distribution on transducer aperture [88], [89]. An excitation signal of arbitrary form can be used and resulting pressure distribution calculated. An example of pressure field distribution for 10 mm diameter circular transducer driven by CW signal at 2 MHz and 5 MHz is presented in Figure 5.18.

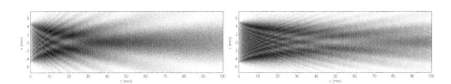

Figure 5.18 Ultrasonic pressure field distribution for planar 10 mm diameter transducer at 2 MHz (left) and 5 MHz (right)

Main disadvantage of planar transducers is existence of near field where no diagnostic measurement can be done because field distribution is similar to plane wave and beam is wide with cylindrical profile. Length of near field zone z_{NF} depends on transducer radius a and on wavelength λ:

$$z_{NF} = \frac{4a^2 - \lambda^2}{4\lambda} \approx \frac{a^2}{\lambda}. \qquad (5.32)$$

Excitation by SS signal causes migration of the z_{NF} along beam axis. Therefore detection of objects near the transducer is complicated. One of the possible solutions is use of focused transducers with acoustical lens or concave piezoelement (Figure 5.19).

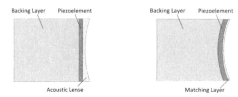

Figure 5.19 Focused ultrasonic transducers with acoustical lens (left) and concave piezoelement (right)

Such transducers are characterized by focal distance F which depends on transducer curvature or acoustical properties and curvature of the lens. The effective focal distance F_e is slightly lower [90]:

$$F_e = \frac{F}{1 + \frac{2}{3}(S_F)^{4/3}},\tag{5.33}$$

where $S_F = F/z_{NF}$. For instance, a circular 10 mm diameter transducer with focal distance $F=20$ mm has $F_e=15$ mm in case of 2 MHz excitation and F_e is 18 mm when operating at 5 MHz. Refer to Figure 5.20 for simulated field for such transducer.

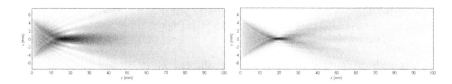

Figure 5.20 Ultrasonic pressure field distribution for focused 10 mm diameter transducer with focal length 20 mm at 2 MHz (left) and 5 MHz (right) excitation

Ultrasonic beam stays focused only at effective focal distance F_e then it starts to diverge. Focal zone length at -6 dB level can be estimated as [91]:

$$F_{Z-6dB} = z_{NF} \cdot S_F{}^2 \cdot \frac{2}{1+0.5 \cdot S_F}.$$
(5.34)

One can see that focal zone length is frequency dependent. For the above mentioned transducers focal zone length is F_Z=18.5 mm for 2 MHz frequency and F_Z=8.6 mm for 5 MHz.

Use of wideband signals varies effective focal distance and it moves closer to transducer if lower frequencies prevail in excitation signal. This is the case for pulse excitation signal (refer to Figures 5.3 and 5.15). For SS signal spectral content can be varied. Usually all the frequency components are distributed evenly (refer to Figures 5.3 and 5.15). Therefore effective focal distance should be more stable with SS excitation. Beam diameter of focused transducer can be estimated as:

$$D_{-6dB} = 1.02\frac{F \cdot c_0}{f \cdot 2a} = 0.5136 \cdot a \cdot S_F.$$
(5.35)

Above mentioned techniques can be used for approximate transducer directivity parameters estimation. Additional problem is caused by wide bandwidth of SS signals [92]. The most accurate results can be obtained using FEA modeling of transducers and propagation [93, 94].

Main advantage of FEA is the possibility to simulate transducers of any shape and acoustic media with any inhomogeneity in it. Transducer and acoustic media are divided into finite elements and wave propagation is calculated in time domain. A grid with at least 10 elements per wavelength element is necessary in order to avoid spatial frequency components aliasing errors. This increases the simulation time.

Example of the wave transmitted by unfocused transducer under pulse and chirp excitation is presented in Figure 5.21. Pulse center frequency was 0.2 MHz, chirp was 15 µs long with frequency variation from 0.3 MHz to 3 MHz. It can be seen that chirp, despite higher frequency range, occupies much larger space in material.

Ringdown artifacts can be seen in the tail of the SS signal even after significant propagation time. Problem gets even worse when focused wave propagation is analyzed. Refer to Figure 5.22 for FEA simulation of the same media and signals as in above example. Concave transducer was used for wave focusing.

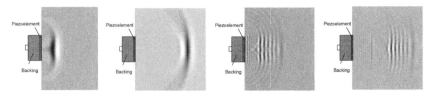

Figure 5.21 Ultrasonic field transmitted by plane circular transducer in case of pulse (left) and chirp (right) excitation at different time instances

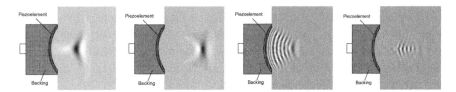

Figure 5.22 Ultrasonic field transmitted by focused transducer in case of pulse (left) and chirp (right) excitation at different time instances

Despite being sufficiently accurate, FEA is not capable to take the real situation into account. Therefore experimental transducer directivity measurement is the reflecting its properties most accurately. Next part is dedicated to directivity measurement techniques.

5.4.2   Directivity measurement techniques

According to [95] directivity is measured registering the probe acoustic beam in water, using a reflecting target. This target usually is a small reflector which can be regarded as a point source, or a hydrophone receiver. The beam parameters are determined by: i) scanning the reflector or hydrophone relative to the transducer beam, or ii) moving the target or the hydrophone.

In case of reflector pulse-echo mode is used: same transducer is used for signal transmission and reception. For this reason beam width is evaluated at -6 dB from peak (2x3 dB). Standard [95] assigns a small steel ball as reflector: for frequencies 0.5 MHz to 3 MHz (3-5) mm diameter is recommended, for (3-15) MHz range ball diameter should be below 3 mm. In our case, SS signals are much longer than pulse: if for 5 MHz transducer optimal pulse duration can be 100 ns, SS signal could be microseconds long. In such case SS signal reflections from ball front face and inner surface will overlap (refer Figure 5.23 for 3 mm ball size comparison with physical signal length is steel).

Standard only briefly mentions the rod use (Table 2 in [95]), but in case of SS signals rod as reflector is a must.

If hydrophone is used, transducer is tested only in transmission mode and only the transmitting directivity of the probe is verified. In such case beam width is evaluated at -3 dB of peak.

Figure 5.23 Reflection of pulse signal from steel ball target is short but chirp is much longer than ball dimensions (left); therefore rod as reflector is preferred in case of SS signals (right)

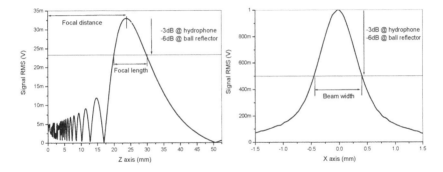

Figure 5.24 Axial beam profile represents the pressure along beam axis (left); transverse beam profile indicates the signal amplitude change across beam profile (right)

Beam measurement can be performed either manually or using automated scanning system. Several beam parameters can be rectified: i) focal distance (distance along beam at which focusing is most efficient so reflection amplitude is maximized) and ii) length of focal zone (range along beam where signal drops -3 dB from peak) are obtained from axial beam profile; iii) -3 dB or -6 dB width of focal zone (beam width) is obtained from transverse beam profile (Figure 5.24); iv) beam divergence angle can be found for focused probes using two adjacent x-y scans on z axis and the corresponding focal width values.

In order to collect full data representing the transducer beam 3D scanning is necessary: x-y scan represents just a single slice of the beam.

5.5 Ultrasonic signals acquisition system

If transducer supposed to be used in imaging, then directivity of the transducer is important. Dedicated data acquisition system is needed for directivity evaluation. In addition, aim here was to explore the imaging capabilities of the SS signals. In order not to limit the SS performance of the system the ultrasonic transducer must match the bandwidth of the excitation signal [96]. Therefore investigation of transducer performance under complex excitation signal [46], [97] was used where the spectral content influence of directivity was studied to compare different signals.

5.5.1   System structure and data processing

Data acquisition system, dedicated for spread spectrum signals, has been designed (Figure 5.25). System composed by: i) high voltage pulse trains generator, located at front end module; ii) signal reception unit located at front end module; iii) filtering, digitization, digital processing and data transfer module; iv) 3D positioning scanner and the corresponding controller. High voltage pulser is capable of conventional and spread spectrum signals excitation [42]. Signal reception unit has a programmable gain. Main acquisition module is located close to host PC, while front-end module is located close to ultrasonic transducer to minimize EMI.

Excitation signals amplitude can be varied from 1 V to 300 V. Excitation signal's rise/fall times of 10 ns ensure 30 MHz system bandwidth. Pulse train duration step is 10 ns. Reception channel's gain can be varied in (0-45) dB range. Analog-to-digit conversion frequency is 100 MHz and uses 10 bit resolution. Minimum scanning step along x and y axes is 10 μm, and 7.5 μm along z axis. System is capable to use several types of signals for experiment simultaneously. This ensures that data obtained for several signals can be reliably compared.

The reflection mode directivity of the transducer was investigated using 5 mm steel bar with a spherical end as point target (Figure 5.25, right). Scan was done along x axis first, then y coordinate was stepped. Same step was used along x and y axes. Slices along z axis were taken using variable step (Figure 5.26, right) to adjust z-slice positions to beam intensity variation along this axis (Figure 5.24). Larger steps were used where variation was slow, while spacing between x-y scan planes was small in areas with large axial pressure variation.

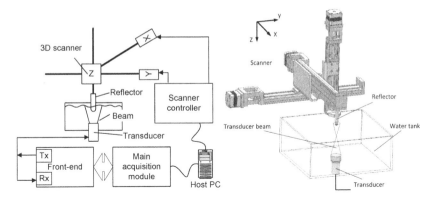

Figure 5.25 Data acquisition system structure for reflection directivity evaluation

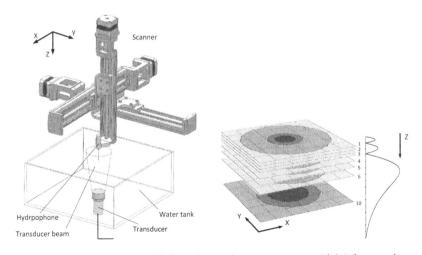

Figure 5.26 Positioning system (left) and scan planes arrangement (right) for transducer pressure field directivity measurement

Collected A-scans were stored for further processing. Application of long SS signals created additional problem: reflection from the steel ball backwall was too close to the front reflection's tail. Therefore, accurate signal gating was needed. Generator waveform length is obtained from computer file used for excitation signal description. Signals acquired at focal point (at distance L_0 from transducer) are taken and location where rms value of the A-scan has a peak established along x and y axes. Peak location x_0 and y_0 coordinates were stored for further processing and corresponding A-scan signal was processed to locate the temporal position for envelope.

Envelope was taken using Hilbert transform:

$$s_{env}(t) = \left| \mathcal{F}^{-1}\left(2 \cdot \mathcal{F}[s(t)]_{=0\,\forall f<0}\right)\right|, \tag{5.36}$$

where $s(t)$ is signal in time domain; $s_{env}(t)$ is the signal envelope obtained; and $\mathcal{F}^{-1}$ and $\mathcal{F}$ are inverse and forward Fourier transforms respectively. Envelope peak position and the length of the excitation signal were used to automatically calculate the signal gating position.

Same procedure was iterated for another measurement plane, obtained at distance L_{FF} from transducer (further than focal point, in far field). Peak coordinates x_{FF} and y_{FF} were stored for further processing. Then, gated signals at focal point and the further position were cross correlated to establish the ToF difference between the two using (5.1). Thanks to high SNR, high accuracy of ToF can be obtained. But sampling period is limiting the ToF estimation resolution. Subsample estimate for signals delay was obtained using cosine interpolation [96]:

$$\Delta ToF_{cos} = -\frac{\theta}{f_s \omega_0}, \tag{5.37}$$

Where f_s is the sampling frequency and

$$\omega_0 = \arccos\left(\frac{x_{m-1} + x_{m+1}}{2x_m}\right), \theta = \arctan\left(\frac{x_{m-1} - x_{m+1}}{2x_m \sin\omega_0}\right). \tag{5.38}$$

Once x-y plane positions were known, and positioning accuracy is high, ToF can be used to obtain the accurate velocity in water, which later can be used for gating position calculation:

$$c_w = \left(\frac{L_{FF} - L_0}{2 \cdot ToF}\right). \tag{5.39}$$

Another problem is associated with transducer's inclination: even accurate positioning of the beam inside the scan did not ensure that succeeding planes are centered. Offset of peak location from center can be obtained by getting the beam center offset values Δx_{FS}, Δy_{FS} obtained at focal spot (z_{FS}) and values Δx_0, Δy_0, obtained at previous z-plane with offset z_0 can be used for inclination coefficients β, x_{01} and y_{01} calculation:

$$\beta_{zx} = \frac{\left(\Delta x_0 - \Delta x_{FS}\right)}{\left(z_0 - z_{FS}\right)}, \beta_{zx} = \frac{\left(\Delta y_0 - \Delta y_{FS}\right)}{\left(z_0 - z_{FS}\right)}, \quad (5.40)$$

$$x_{01} = \frac{\left(\Delta x_0 \cdot z_{FS} - \Delta x_{FS} \cdot z_0\right)}{\left(z_{FS} - z_0\right)}, y_{01} = \frac{\left(\Delta y_0 \cdot z_{FS} - \Delta y_{FS} \cdot z_0\right)}{\left(z_{FS} - z_0\right)}. \quad (5.41)$$

Then beam inclination compensation for every z-slice is:

$$x_{offset_i} = x_{01} + \beta_{zx} z_i - x_{center}, y_{offset_i} = y_{01} + \beta_{zy} z_i - y_{center}, \quad (5.42)$$

where z_i is the z-slice distance and x_{center} and y_{center} are desired x and y center locations after beam alignment. Desired compensation can be applied via circular shift of x-y plane data.

5.5.2 Directivity investigation results

Pressure field directivity study [17] was done for flat beam wideband composite transducer TF5C6N-E supplied by Doppler. Transducer was placed in the bottom wall of the water tank (Figure 5.26). Excitation was using 4.96 MHz CW burst signal of 3 μs duration and bipolar, +/30 V (60 Vpp) amplitude. Pressure was measured by scanning the hydrophone HNP1000 from ONDA at desired distances (defined by scan plan) from transducer over (50x50) mm xy range, using 0.2 mm step. Signal RMS is obtained at every z-axis location of x-y beam profile. Obtained pressure field distribution at near field and far field is presented in Figure 5.27.

Directivity diagram is almost flat at small distances (Figure 5.27, left) and converges to Gaussian form at far field (Figure 5.27, right). Directivity evaluation in reflection mode used three types of excitation signals:
i) single pulse as representative of classical wideband excitation;
ii) continuous wave (CW) burst as classical narrowband excitation;
iii) chirp, as SS signal representative.

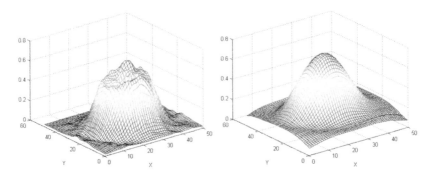

Figure 5.27 Transducer pressure field obtained using hydrophone for flat beam transducer
TF5C6N-E at 6 mm (left) and 62 mm (right) distance from transducer face

Full beam profile of wideband lens focused 10 MHz transducer
IRY210 was collected using round end bar as reflector. Signals were
collected using a 50 ns duration pulse (optimal for 10 MHz), 1 μs duration
CW burst 10 MHz, and the 1 μs duration chirp with frequencies span (2.5-
20) MHz. Results are presented in Figure 5.28.

It can be seen that CW toneburst has the largest energy. Chirp energy is
lower since some of frequency components were spread over higher
attenuation frequencies. Pulse signal energy is the lowest. Normalizing to
peak value reveals other differences: pulse signal has least variation along
axis; focal distance is different for all signals.

In order to evaluate the beam width, signal variation along z axis can
be removed and every z-slice analyzed for -3 dB, -6 dB and -20 dB signal
drop to obtain the beam width at every z axis position. Beam profile
analysis results same transducer IRY210 are presented in Figure 5.29.

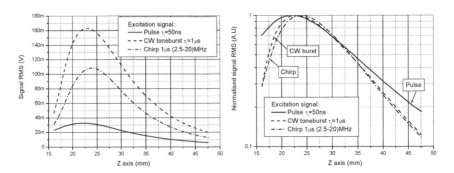

Figure 5.28 Axial beam profile along z axis for transducer IRY210 when excited by pulse
CW burst and SS signal: absolute values (left) and normalized to peak (right)

For pulse excitation, beam width is larger than in case of SS signal. Same can be found from transversal beam profile obtained at 22 mm distance (focal length for pulse signal) presented in Figure 5.30. Beam width differences are clear: pulse excitation produces 0.85 mm width, CW burst is 0.7 mm and chirp has 0.6 mm beam at -6 dB level.

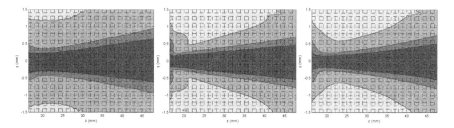

Figure 5.29 Axial beam profile z-y slice for transducer IRY210 when excited by pulse (left), CW burst (center) and SS signal (right)

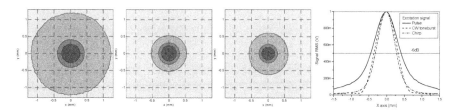

Figure 5.30 Transducer IRY210 transversal beam profile (3 dB, 6 dB and 20 dB) x-y slice at focal spot when excited by (left to right) pulse, CW burst and SS signal; x-plot for all (right)

Explanation of spectral content influence can be seen in Figure 5.31: beam width for the every spectral component is calculated separately here.

It can be seen that beam size here is similar. But pulse has more low frequency components: this can be seen from Figure 5.3. Lower frequency components form wider beam. Therefore resulting beam is wider. Energy is spread more evenly for chirp therefore contribution from all frequencies is uniform. CW toneburst has all the energy concentrated at transducer nominal frequency therefore moderate beam performance is observed.

If normalization is used along scanning direction, spectral content change with spatial position can be studied (Figure 5.32).

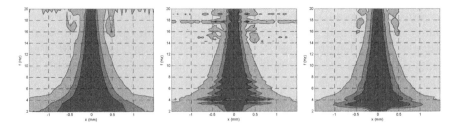

Figure 5.31 Transversal beam profile (3 dB, 6 dB and 20 dB) for IRY210 transducer over frequency range covered by excitation signal (left to right): pulse, CW burst and SS signal

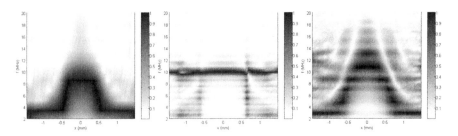

Figure 5.32 Excitation signals spectrum: pulse (left), CW burst (center) and chirp (right)

It can be seen that SS signal produces better transducer bandwidth coverage, especially in high frequency domain. Pulse signal has more low frequency content. The CW burst has high SNR, but all energy is concentrated at 10 MHz. CW burst is able to maintain the 10 MHz content at high offsets from axis, while pulse and SS signals have only low frequencies. Conclusion can be drawn that SNR at high frequencies is higher for SS and SNR for low frequencies is better for pulse. Meanwhile, the CW burst signal has acceptable SNR only around 10 MHz.

5.6 Summary

This chapter has provided the ultrasonic data acquisition principles and techniques used for the ultrasonic transducer properties investigation under spread spectrum excitation. Spread spectrum signals properties were discussed and differences from conventional signals outlined. Importance of sharp correlation mainlobe and low sidelobes in time of flight estimation was indicated. Discussion on signal types presented main types of spread spectrum signals. Transducer type discussion indicated essential transducer parameters and techniques for these parameters estimation. Electrical energy transduction into pressure investigation was presented. It was shown that transducer and excitation generator impedance interact so it is important to acquire both. Techniques for impedance evaluation in low power CW probing and high power pulsed and SS probing have been presented. It can be seen that despite being able to deliver more energy to the wider bandwidth spread spectrum signals usually produce spectrum which is not smooth. Therefore spectral interpretation is more complicated compared to the one produced by pulse signal. Yet, SS signals do not have the spectral zeros as it is the case with pulse signal therefore probing is more stable. Setup and system structure are outlined for directivity investigation in reflection and pressure field mode. Evaluation of the ultrasonic transducer performance under excitation by spread spectrum signals requires specific excitation and reception equipment. It was indicated that SS signals put certain limitations on acquisition equipment: excitation and reception equipment should account for signal specifics. Excitation unit should be able to deliver the power required and be capable to store in memory and reproduce the SS signals. Since spread spectrum signals are significantly longer than conventional, use of ball as reflecting target is not suitable since reflections from ball front and back surface will overlap. It was suggested to replace the ball with rod with spherical end. Directional relation of spectral content to signal type and off-axis distance is presented. Comparison of experimentally obtained transducer directivity indicate that SS probed spectral content is different from that obtained using conventional pulse and toneburst signals. Investigation revealed that SS signals are able to deliver more energy into high frequency range. Thanks to this property their directivity is better than in case of pulse or CW burst excitation.

References

[1] J.D.N. Cheeke. *Fundamentals and Applications of Ultrasonic Waves, Second Edition*. CRC press, 2012.

[2] S. Dixon, C. Edwards, and S.B. Palmer. High accuracy non-contact ultrasonic thickness gauging of aluminium sheet using electromagnetic acoustic transducers. *Ultrasonics*, 39(6):445-453, 2001.

[3] A.N. Kalashnikov, V. Ivchenko, R.E. Challis, and A.K. Holmes. Compensation for temperature variation in ultrasonic chemical process monitoring. In *Ultrasonics Symposium, 2005 IEEE*, 2, 2005:1151-1154.

[4] M. Vázquez, L. Leija, A. Vera, A. Ramos, and E. Moreno. Experimental Estimation of Acoustic Attenuation and Dispersion. In *ICEEE and CIE 2005*, 2005:156-159.

[5] B.R. Shinde, S.S. Jadhav, S.U. Shinde, D.R. Shengule, and K.M. Jadhav. Ion-solvent interactions studies in aqueous manganous chloride solution by ultrasonic velocity measurement at different temperatures. *Archives of Physics Research*, 2(2):107-113, 2011.

[6] A. Afaneh, S. Alzebda, V. Ivchenko, and A.N. Kalashnikov. Ultrasonic Measurements of Temperature in Aqueous Solutions: Why and How. *Physics Research International*, 2011:1-10, 2011.

[7] M.M. Saad, C.J. Bleakley, T. Ballal, and S. Dobson. High-accuracy reference-free ultrasonic location estimation. *IEEE Trans. Instrum. Meas.*, 61:1561-1570, 2012.

[8] M. Hazas and A. Hopper. Broadband ultrasonic location systems for improved indoor positioning. *IEEE Trans. Mobile Comput*, 5(5):536-547, 2006.

[9] A. Cuerva and A. Sanz-Andres. On sonic anemometer measurement theory. *Journal of Wind Engineering and Industrial Aerodynamics*, 88:25-55, 2000.

[10] Q. Licun and R.L. Wang. Performance improvement of ultrasonic Doppler flowmeter using spread spectrum technique. In *IEEE International Conference on Information Acquisition*, 2006:122-126.

[11] R. Righetti, J. Ophir, and P. Ktonas. Axial resolution in elastography. *Ultrasound in medicine and biology*, 28(1):101-113, 2002.

[12] A. Voleisis et al. Ultrasonic method for the whole blood coagulation analysis. *Ultrasonics*, 40(1-8):101-107, May 2002.

[13] C. Fritsch et al. A Full Featured Ultrasound NDE System in aStandard FPGA. In *9th European Conference on NDT*, Berlin, 2006:1-10.

[14] S. Wagle and H. Kato. Real-time Measurement of Ultrasonic Waves at Bolted Joints under Fatigue Testing. *Experimental Mechanics*, 51:1559-1564, 2011.

[15] I. Ihara. Ultrasonic Sensing: Fundamentals and its Applications to Nondestructive Evaluation. In *Sensors*, S. C. Mukhopadhyay and R. Y. M. Huang (Eds.)., Springer Berlin Heidelberg, 2008, 21:287-305.

[16] C. Rao. Information and the accuracy attainable in the estimation of statistical parameters. *Bulletin of Calcutta Mathematics Society*, 37:81-89, 1945.

[17] L. Svilainis, V. Dumbrava, A. Chaziachmetovas, D. Kybartas, and V. Eidukynas. Investigation of the ultrasonic transducer suitability for spread spectrum systems. In *Intelligent Data Acquisition and Advanced Computing Systems (IDAACS), 2013 IEEE 7th International Conference on*, 01, 2013:41-46.

[18] T. Gang, Z.Y. Sheng, and W.L. Tian. Time resolution improvement of ultrasonic TOFD testing by pulse compression technique. *INSIGHT*, 54:193-197, 2012.

[19] S. Olcum, M.N. Senlik, and A. Atalar. Optimization of the gain-bandwidth product of capacitive micromachined ultrasonic transducers. *IEEE Trans. Ultrason., Ferroelect., Freq. Contr.*, 52(12):2211-2219, 2005.

[20] L. Svilainis. Review of high resolution time of flight estimation techniques for ultrasonic signals. In *BINDT-NDT*, Telford, 2013:1-12.

[21] L.R. Welch and M.D. Fox. Practical spread spectrum pulse compression for ultrasonic tissue imaging. *IEEE transactions on ultrasonics ferroelectrics and frequency control*, 45:349-355, 1998.

[22] T. Virolainen, J. Eskelinen, and E. Haeggstrom. Frequency domain low time-bandwidth product chirp synthesis for pulse compression side lobe reduction. In *Ultrasonics Symposium (IUS), 2009 IEEE International*, 2009:1526-1528.

[23] M. Zapf, B.F. Derouiche, and N.V. Ruiter. Evaluation of chirp and binary code based excitation pulses for 3D USCT. In *Ultrasonics Symposium (IUS), 2009 IEEE International*, 2009:1996-1999.

[24] S.W. Huang and P.C. Li. Arbitrary waveform coded excitation using bipolar square wave pulsers in medical ultrasound. *IEEE Trans. Ultrason. Ferroelect. Freq. Contr.*, 53:106-116, 2006.

[25] L. Svilainis, V. Dumbrava, S. Kitov, and A. Chaziachmetovas. The influence of digital domain on time of flight estimation performance. In *in: Proc. International congress on ultrasonics, AIP Conference Proceedings*, 2012:122-126.

[26] J.E. Michaels, S.J. Lee, A. Croxford, and P.D. Wilcox. Chirp excitation of ultrasonic guided waves. *Ultrasonics*, 53:265-270, 2013.

[27] J. Park, C. Hu, X.L. Xiang, Q. Zhou, and K.K. Shung. Wideband Linear Power Amplifier for High-Frequency Ultrasonic Coded Excitation Imaging. *IEEE transactions on ultrasonics ferroelectrics and frequency control*, 825-832:59, 2012.

[28] M.A. K. Afzal, S.F. Russell, J.K. Kayani, and S.J. Wormley. Application of Spread-Spectrum Ultrasonic Evaluation to Concrete Structures. In *Review of Progress in Quantitative Nondestructive Evaluation*, D. Thompson and D. Chimenti (Eds.)., Springer US, 1996:1771-1778.

[29] M. Pollakowski and H. Ermert. Chirp signal matching and signal power optimization in pulse-echo mode ultrasonic nondestructive testing. *IEEE Trans. Ultrason., Ferroelect., Freq. Contr.*, 41(5):655-659, 1994.

[30] G. Athanasopoulo, C. Stephen, and J. Hatfield. A high voltage pulser ASIC for driving high frequency ultrasonic arrays. In *Ultrasonics Symposium, 2004 IEEE*, 2, 2004:1398-1400.

[31] L. Svilainis, K. Lukoseviciute, V. Dumbrava, A. Chaziachmetovas, and A. Aleksandrovas. Application of Arbitrary Position and Width Pulse Trains Signals in Ultrasonic Imaging: Correlation Performance Study. *Electronics and Electrical Engineering*, 19(3):57-60, 2013.

[32] J. Minkoff. *Signal Processing Fundamentals and Applications for Communications and Sensing Systems*. London, Artech House, 2002.

[33] S. Bjorklund. A Survey and Comparison of Time-Delay Estimation Methods in Linear Systems. Ph.D. dissertation Linkoping university, 2003.

[34] G. Jacovitti and G. Scarano. Discrete time techniques for time delay estimation. *IEEE Trans. Signal Processing*, 41:525-533, 1993.

[35] R. Demirli and J. Saniie. Model-based estimation of ultrasonic echoes. Part I: Analysis and algorithms. *IEEE Trans. Ultrason., Ferroelect., Freq. Contr.*, 48(3):787-802, 2001.

[36] L. Zhang and X. Wu. On the application of cross correlation function to subsample discrete time delay estimation. *Dig. Signal Processing*, 16:682-694, 2006.

[37] H. Cramer. *Mathematical Methods of Statistics*. Princeton University Press, 1999.

[38] L. Svilainis, V. Dumbrava, and D. Kybartas. Evaluation of the ultrasonic preamplifier noise voltage density. *Journal of Circuits, Systems, and Computers*:In press, 2013.

[39] A. Rodríguez, A. Salazar, and L. Vergara. Analysis of split-spectrum algorithms in an automatic detection framework. *Signal Processing*, 92(9):2293-2307, September 2012.

[40] L. Svilainis et al. Comparison of spread spectrum and pulse signal excitation for split spectrum techniques composite imaging. In *IOP Conference Series: Materials Science and Engineering*, 42, 2012:012007.1-4.

[41] L. Svilainis, A. Chaziachmetovas, and V. Dumbrava. Efficient high voltage pulser for piezoelectric air coupled transducer. *Ultrasonics*, 53:225-231, 2013.

[42] L. Svilainis, V. Dumbrava, A. Chaziachmetovas, and A. Aleksandrovas. Pulser for arbitrary width and position square pulse trains generation. In *Ultrasonics Symposium (IUS), 2012 IEEE International*, Dresden, 2012:1-4.

[43] Z.-J. Yao, Q.-H. Meng, G.-W. Li, and P. Lin. Non-crosstalk real-time ultrasonic range system with optimized chaotic pulse position-width modulation excitation. In *Ultrasonics Symposium, 2008. IUS 2008. IEEE*, 2008:729-732.

[44] L. Fortuna, M. Frasca, and A. Rizzo. Chaotic pulse position modulation to improve the efficiency of sonar sensors. *IEEE Trans Instrum Meas*, 52(6):1809-1814, 2003.

[45] U. Grimaldi and M. Parvis. Enhancing ultrasonic sensor performance by optimization of the driving signal. *Measurement* , 14:219-228, 1995.

[46] C. Papageorgiou, C. Kosmatopoulos, and T. Laopoulos. Automated characterization and calibration of ultrasonic transducers. In *MELECON 98*, 2, 1998:1214-1218.

[47] S.-H. Chang, N.N. Rogacheva, and C.C. Chou. Analysis of methods for determining electromechanical coupling coefficients of piezoelectric elements. *IEEE Trans. Ultrason., Ferroelect., Freq. Contr.*, 42(4):630-640, 1995.

[48] R. Kazys et al. High temperature ultrasonic transducers for imaging and measurements in a liquid Pb/Bi eutectic alloy. *IEEE Trans. Ultrason., Ferroelect., Freq. Contr.*, 52(4):525-537, 2005.

[49] O. Oralkan et al. Capacitive micromachined ultrasonic transducers: next-generation arrays for acoustic imaging? *IEEE Trans. Ultrason., Ferroelect., Freq. Contr.*, 49(11):1596-1610, 2002.

[50] Y.-M.R. Huang (Ed.). *Sensors. Advancements in Modeling, Design Issues, Fabrication and Practical Applications*. Springer US, 2008.

[51] W. Heywang, K. Lubitz, and W. Wersing (Eds.). *Piezoelectricity. Evolution and Future of a Technology*. Sp, 2008.

[52] A. Mohammadi, M.H. Jafari, and A. Khanafari. A procedure to extend the performance of optimization methods in ultrasonic transducer matching layer design. In *Fuzzy Systems (IFSC), 2013 13th Iranian Conference on*, 2013:1-5.

[53] P. Reynolds, J. Hyslop, and G. Hayward. Resonant characteristics of piezoelectric composites: analysis of spurious modes in single and multi-element ultrasonic transducers. In *Ultrasonics Symposium, 2002. Proceedings. 2002 IEEE*, 2, 2002:1157-1160.

[54] A. Perry, C.R. Bowen, H. Kara, and S. Mahon. Modelling of 3-3 piezocomposites. *Software for Electrical Engineering Analysis and Design V. Engineering Sciences*, 31, 2001, Wessex Institute Of Technology, United Kingdom.

[55] L. Capineri, L. Masotti, M. Rinieri, and S. Rocchi. Ultrasonic transducers as a black-box: equivalent circuit synthesis and matching network design. *IEEE Trans. Ultrason., Ferroelect., Freq. Contr.*, 40(6):694-703, 1993.

[56] R.E. McKeighen. Design guidelines for medical ultrasonic arrays. In *Medical Imaging 1998: Ultrasonic Transducer Engineering*, 2, 3341, May 1 1998:2-18.

[57] T. Gudra and K.J. Opielinski. Influence of acoustic impedance of multilayer acoustic systems on the transfer function of ultrasonic airborne transducers. *Ultrasonics*, 40:457-463, 2002.

[58] R. Krimholtz, D.A. Leedom, and G.L. Matthaei. New equivalent circuits for elementary piezoelectric transducers. *Electron. Lett*, 6(13):398-399, 1970.

[59] M. Castillo, P. Acevedo, and E. Moreno. KLM model for lossy piezoelectric transducers. *Ultrasonics* , 41(8):671-679, 2003.

[60] M. Prokic. *Piezoelectric Transducers Modeling and Characterization*. M.P. Interconsulting, 2004.

[61] R. McKeighen. Finite element simulation and modeling of 2-D arrays for 3-D ultrasonic imaging. *IEEE Trans. Ultrason., Ferroelect., Freq. Contr.*, 48(5):1395-1405, 2001.

[62] G. Wojcik et al. Computer modeling of diced matching layers. In *Ultrasonics Symposium, 1996. Proceedings., 1996 IEEE*, 2, 1996:1503-1508.

[63] W. Qi and W. Cao. Finite element study on 1-D array transducer design. *IEEE Trans. Ultrason., Ferroelect., Freq. Contr.*, 47(4):949-955, 2000.

[64] S. Sanchez, M. de Espinosa F.R., and N. Lamberti. Multifrequency piezoelectric composites: one-dimensional modeling. *Ultrasonics*, 37:97-105, 1999.

[65] W.P. Mason. *Electromechanical transducers and wave filters*. New York, D. Van Nostrand Co., 1948.

[66] S. Sherrit, S.P. Leary, B.P. Dolgin, and Y. Bar-Cohen. Comparison of the Mason and KLM equivalent circuits for piezoelectric resonators in the thickness mode. In *Ultrasonics Symposium, 1999. Proceedings. 1999 IEEE*, 2, 1999:921-926.

[67] T.L. Rhyne. Characterizing ultrasonic transducers using radiation efficiency and reception noise figure. *IEEE Trans. Ultrason., Ferroelect., Freq. Contr.*, 45(3):559-566, 1998.

[68] V.N. Khmelev, I.I. Savin, R.V. Barsukov, and S.N. Tsyganok. Problems of electrical matching of electronic ultrasound frequency generators and electroacoustical transducers for ultrasound technological installations. In *Electron Devices and Materials, 2004. Proceedings. 5th Annual. 2004 International Siberian Workshop on*, 2004:211-215.

[69] *Impedance Measurement Handbook.* Agilent Technologies Co. Ltd., 2003.

[70] G.L. Amorese. Reducing ESR Measurement Errors. *Microwaves and RF*:75-80, 2003.

[71] J. Hoja and G. Lentka. An analysis of a measurement probe for a high impedance spectroscopy analyzer. *Measurement*, 41:65-75, 2008.

[72] V. Dumbrava and L. Svilainis. The Automated Complex Impedance Measurement system. *Electronics and electrical engineering*, 76(4):59-62, 2007.

[73] L. Svilainis and V. Dumbrava. Test Fixture Compensation Techniques in Impedance Analysis. *Electronics and electrical engineering*, 87(7):37-40, 2008.

[74] V. Juska, L. Svilainis, and V. Dumbrava. Analysis of piezomotor driver for laser beam deflection. *Journal of Vibroengineering*, 11(1):17-26, 2009.

[75] G.J. Lewis, G.S. Lewis, and W. Olbricht. Cost-effective broad-band electrical impedance spectroscopy measurement circuit and signal analysis for piezo-materials and ultrasound transducers. *Meas Sci Technol.*, 19(10):1-7, 2008.

[76] A. Petošic, M. Budimir, and N. Pavlovic. Comparison between piezoelectric material properties obtained by using low-voltage magnitude frequency sweeping and high-level short impulse signals. *Ultrasonics*, 53:1192-1199, 2013.

[77] T. Sun, S. Gawad, C. Bernabini, N.G. Green, and H. Morgan. Broadband single cell impedance spectroscopy using maximum length sequences: theoretical analysis and practical considerations. *Meas. Sci. Technol.*, 18:2859-2868, 2007.

[78] J. Emeterio, A. Ramos, P. Sanz, and A. Ruiz. Evaluation of Impedance Matching Schemes for Pulse-Echo Ultrasonic Piezoelectric Transducers. *Ferroelectrics*, 273:297-302, 2002.

[79] L. Svilainis and V. Dumbrava. Evaluation of the ultrasonic transducer electrical matching performance. *Ultrasound*, 62(4):16-21, 2007.

[80] G.L. Petersen. Matching the Output of a RITEC Gated Amplifier to an Arbitrary Load. USA, 2006.

[81] R.W. Brounley. Matching Networks for Power Amplifiers Operating into High VSWR Loads. *HF Electronics*:58-62, 2004.

[82] M. Garcia-Rodriguez et al. Low cost matching network for ultrasonic transducers. *Physics Procedia*, 3(1):1025-1031, 2010, International Congress on Ultrasonics, Santiago de Chile, January 2009.

[83] H. Huang and D. Paramo. Broadband electrical impedance matching for piezoelectric ultrasound transducers. *IEEE Trans. Ultrason., Ferroelect., Freq. Contr.*, 58(12):2699-2707, 2011.

[84] A. Sertbas and B.S. Yarman. A Computer-Aided Design Technique for Lossless Matching Networks with Mixed, Lumped and Distributed Elements. *Int. J. Electron. Commun.*, 58:424-428, 2004.

[85] *Non-destructive testing - Characterization and verification of ultrasonic examination equipment - Part 1: Instruments.* CEN standard EN12668-1, 2010.

[86] Converting from EOC sensitivity to amplifier-loaded sensitivity. Onda Corp., Sunnyvale, CA, USA, 2007.

[87] S. Leeman et al. Measurement of transducer directivity function. In M.F. Insana and K.K. Shung (Eds.), *Medical Imaging 2001: Ultrasonic Imaging and Signal Processing*, 4325, 2001:47-53.

[88] S. Holm. Simulation of acoustic fields from medical ultrasound tranducers of arbitrary shape. In *Proceedings of Nordic Symposium in Physical Acoustics*, 1995.

[89] J.A. Jensen and N.B. Svendsen. Calculation of pressure fields from arbitrarily shaped, apodized, and excited ultrasound transducers. *IEEE Trans. Ultrason., Ferroelect., Freq. Contr.*, 39(2):262-267, 1992.

[90] P. Marechal, F. Levassort, L.-P. Tran-Huu-Hue, M. Lethiecq, and N. Felix. Effect of acoustical properties of a lens on the pulse-echo response of a single element transducer. In *Ultrasonics Symposium, 2004 IEEE*, 3, 2004:1651-1654.

[91] G. Kossoff. Analysis of focusing action of spherically curved transducers. *Ultrasound in Medicine & Biology*, 5(4):359-365, 1979.

[92] S. Leeman, A.J. Healey, and J.P. Weight. A New Approach for Calculating Wideband Fields. In *Acoustical Imaging*, H. Lee (Ed.)., Springer US, 2002, 24:49-56.

[93] J. Assaad, A.-C. Hladky, and B. Cugnet. Application of the FEM and the BEM to compute the field of a transducer mounted in a rigid baffle (3D case). *Ultrasonics*, 42(1-9):443-446, April 2004.

[94] G.L. Wojcik, D.K. Vaughan, V. Murray, and J. .J. Mould. Time-domain modeling of composite arrays for underwater imaging. In *Ultrasonics Symposium, 1994. Proceedings., 1994 IEEE*, 2, 1994:1027-1032.

[95] *Non-destructive testing - Characterization and verification of ultrasonic examination equipment - Part 2: Probes*. CEN standard EN 12668-2:2010, 2010.

[96] L. Svilainis, K. Lukoseviciute, V. Dumbrava, and A. Chaziachmetovas. Subsample interpolation bias error in time of flight estimation by direct correlation in digital domain. *Measurement* , 46(10):3950-3958, 2013.

[97] A. Nowicki et al. Comparison of sound fields generated by different coded excitations – Experimental results. *Ultrasonics*, 44(1):121–129, January 2006.

Biographies

Linas Svilainis acquired a PhD degree in 1996 from the Kaunas University of Technology (Lithuania). Since 2009 he is Full time Professor at the Department of Signal Processing, Kaunas University of Technology, Lithuania. His current areas of research are: ultrasound electronics and signal processing; LED video displays research and development; Electromagnetic compatibility and EMI protection of electronic systems.

Andrius Chaziachmetovas acquired a PhD degree in 2007 from the Kaunas University of Technology (Lithuania). Since 2010 – Associated Professor at the Department of Signal Processing, Kaunas University of Technology, Kaunas, Lithuania. His current areas of research are ultrasound electronics design and development, LED dimming, high speed photonics and measurement systems electronics.

Darius Kybartas acquired a PhD degree in 2004 from the Kaunas University of Technology (Lithuania). Since 2011 – Associated Professor at the Department of Signal Processing, Kaunas University of Technology, Kaunas, Lithuania. His current areas of research are ultrasonic transducers electrical parameters measurement and modeling, high frequency electronics and antenna design and development.

Valdas Eidukynas acquired a PhD degree in 1996 from the Kaunas University of Technology (Lithuania). Since 1997 he is Associate Professor at the Department of Engineering Mechanics, Kaunas University of Technology, Lithuania. His current areas of research are: biomechanics; structural safety; mechanical processes simulation.

6

Optimal information fusion in stochastic unknown input observers network

Leonid M. Lyubchyk

Faculty of Informatics and Control
National Technical University "Kharkiv Polytechnic Institute"

Abstract

The problem of optimal information fusion in stochastic unknown-input observer network, with ensure the state and input signal estimation for dynamic systems under the presence of stochastic disturbance and random measurement noise is considered. Stochastic unknown-input observer and observer-based inverse model optimal design method is proposed and the conditions of steady-state observer existence are established. The solution of optimal state estimates fusion problem is presented for distributed stochastic observers and consensus unknown-input observer network.

Keywords: decentralized estimation, information fusion, inverse model, consensus observer, unknown-input observer, sensor network

6.1 Introduction

The problem of the fusion of data provided by sensors, obtained information from different measurements, is of great theoretical and practical interest and has been intensive studied in recent decades. Multisensor data and information fusion methods finds a various application in military and industry in relation with the problems of multivariable and large-scale systems state estimation and control, signal

V. Haasz and K. Madani (Eds.), Advanced Data Acquisition and Intelligent Data Processing, 127–158.

processing, fault detection and processes monitoring when the measurement information provided by sensor network [1–3].

Data fusion techniques combine data from multiple distributed sensors and related information from associated data processing units to ensure improved information system accuracy in comparison than could be achieved by the use of a single sensor alone. Another advantage gained by using multiple sensors fusion is improved observability properties of complex measurement system which combines information from different sources and sensors. Moreover, the information data processing system reliability and fault tolerance may be also improved.

The term data fusion is usually used for raw data obtained directly from the sensors and the term information fusion is employed to define already processed data obtained from data processing units.

In practice are widely used several types of architecture schemes of the fusion the information obtained from the network of sensors or information processing nodes, each with its own processing facility [4–6].

Centralized Architecture implies that the fusion node receives the information from all of the input sources. Therefore, all of the fusion processes are executed in a central processor node that uses the provided raw measurements from the sources or the results of their preprocessing. In this scheme, the sources obtain only the observation as measurements and transmit them together with local estimates to a central processor, where the data fusion process is performed.

Decentralized Architecture is composed of nodes in which each one has its own processing capabilities and there is no single point of data fusion. Data fusion is performed autonomously, with each node processes its local information and the information received from neighboring nodes.

Distributed Architecture involves that measurements from each source node are processed independently before the information is sent to the fusion node; the fusion node uses for the information that is received from the other nodes. In other words, the data processing is performed in the source node before the information is communicated to the fusion node.

Hierarchical Architecture comprises a combination of decentralized and distributed sensors and observers nodes, generating hierarchical schemes in which the data and information fusion process is performed at different levels in the hierarchy.

In principle, decentralized data fusion system architecture is more difficult to implement because of the computation and communication requirements. However, in practice, there is no single best data and information fusion architecture, and the selection of the most appropriate architecture should be made depending on the system requirements, demand, sensor networks properties, measurement data availability, nodes processing capabilities and so on.

One of the most important problems in the field of data and information fusion is state and input signal estimation of large-scale dynamic system by the set of decentralized observers or filters [7–9]. Such a decentralized data processing system may be considered as a Multi-Agent Network (MAN), where each agent represents a separate dynamic observer or filter which estimates the state of whole system, but uses only the local measurement information obtained from the appropriate set of neighboring sensors [10, 11]. The fusion of estimates obtained by separate agents-observers allows significantly improving the estimation accuracy and reducing the amount and difficulties of computation.

In a centralized estimate fusion scheme, each agent-observer sends its estimation data either directly to a data fusion center. The fusion center in turn generates the final summary aggregate estimate. In contrast in fully decentralized fusion scheme there is no central fusion center and any agent-observer exchanges data with its neighbors and carries out local computation. In such a system, fusion occurs locally at each node on the basis of local observations and the information communicated from neighboring nodes. Therefore, each node provides an estimate of the object state based on only their local views, and this information is the input to the fusion process, which provides a fused global view.

Here we will focus on a decentralized state and input signal estimation problem for a distributed system with multiple agents-observers, where each agent estimates the state of the whole system. The problem of decentralized estimation based on distribute filtering has been intensive studied in recent years [12, 13]. A decentralized state estimation data fusion system consists of a network of sensor and observer nodes, each with its own processing facility, which together do not require any central fusion or central communication facility. At that, fusion occurs locally at each observer node on the basis of local observations and the information communicated from neighboring nodes.

Such an approach is based on special structure of decentralized filter where the information from local estimators can yield the global optimal state estimate according to certain information fusion criterion [14, 15]. The main idea of estimates fusion is to use a multi-sensor optimal information fusion criterion in the linear minimum variance sense, which for linear systems with Gaussian noise is equivalent to the maximum likelihood method. Based on this optimal fusion criterion, using the average estimates, weighted by matrices, a general multi-sensor optimal information fusion decentralized Kalman filter has a two-layer fusion structure. First fusion layer has a parallel structure to determine the estimation error covariance matrices and cross covariance between every pair of filters at each time step, which are used for local state estimation. The second fusion layer is the fusion center that determines the optimal fusion matrix weights and obtains the optimal fusion filter.

A significant improvement of decentralized filters theory was the creation of distributed consensus filtering and estimation concept [10, 11, 16, 17]. In recent years was proposed a scalable sensor fusion scheme that requires fusion of sensor measurements combined with local Kalman filtering [18]. The key component of this approach is to develop a distributed algorithm that allows the local filters in the nodes of a sensor network to track the average of all measurements available for such a node and obtained from neighboring nodes. This problem was referring to as dynamic average consensus and the solution was obtained in the form of a distributed low-pass filter, called the *Consensus Filter*, which solves this tracking problem via reaching an average-consensus [17, 18]. The role of consensus filter is to perform distributed fusion of sensor measurements that is necessary for implementation of a scalable Kalman filtering scheme.

Further development of this concept led to the development of distributed consensus Kalman filters includes an additional term formed in accordance with the so-called *Consensus Protocol.* Typically, as this term is usually taken the difference between the current local estimates and weighted average estimates of filters in neighboring network nodes. At that, at each step, every filter-node can compute a local weighted least-squares estimate, which converges to the global optimal solution of state estimate.

Note that recently distributed consensus filters and observers are widely used in multi-agent control systems that implement the coordinative consensus control of large-scale dynamic systems under incomplete measurements as well as control of collective objects behavior and motion.

In this regard, a great interest is a problem of decentralized state estimation of the dynamic system with unknown and immeasurable input signals, as well as unknown input signal estimation. Such a general problem is directly related with important problems of signal blind separation [19], fault detection and isolation [20], disturbance rejection in multivariable and large-scale control systems [21, 22].

The most effective approach to solving such problems is based on the theory of *Unknown-Input Observers* (UIO) [23, 24]. The classical scheme of state estimation of dynamic systems based on Luenberger observers or

Kalman filters require an exact reference or measurement of input signals. In contrast to the classical approach, UIO provide state estimation in the absence of information about the input signal. This is achieved by a special choice of observer parameter matrices in order to ensure the independence of the estimation error from the unknown input signal [23, 24]. This, however, to some extent, restricts of the observer parametric synthesis possibility, which requires for its realization fulfillment of invertability and strong observability conditions of the system. To overcome these difficulties, the special regularization technique may be used [25, 26].

The state-space realization of inverse model with desired dynamic properties can be obtained using UIO technique because observer equation combined with the unknown input signal estimate may be treated as the designed inverse model. In such a way the inverse model design problem includes *structural synthesis* renders the fixed and free parameters and *parametric design* in order to ensure the inverse model stability and desired dynamic properties based on the appropriate parameters tuning methods, such as pole-placement or performance optimization.

The problems of optimal stochastic UIO design and UIO-based inverse systems were considered in [27, 28].

Despite the fact, that the methods of UIO design developed far enough, the problem of decentralized UIO synthesis and fusion technique was considered in relatively small number of publication [29, 30].

Thus, the extension of the methods of states and inputs of stochastic systems invariant estimation based on UIO and inverse models approach for decentralized estimation problem using estimates fusion and consensus design technique is of a significant scientific and practical interest.

The Chapter is organized as following. In Section 2 considered the problem statement of optimal data fusion in the network of stochastic UIO, with ensure the state and input signal estimation for dynamic systems under the presence of stochastic disturbance and random measurement noise. The problem stated for two main data fusion system architecture, namely, centralized estimation fusion scheme with estimates aggregation and decentralized consensus observers fusion scheme.

In Section 3 the solution of state and input signal deterministic estimation problem under conditions of absence of stochastic disturbance and noise is examined. At first, the structural and parametric design procedure for reduced-order UIO and UIO-based inverse model is presented. For the case when the solvability conditions for the parametric synthesis of the UIO do not met, the suitable regularization procedure is used. The solution of identification problem of dynamical systems unknown input signals is also considered for signals which characterized by a wave structure, at that the input estimates produced by UIO based inverse model are used as *on-line* information for the identification procedure.

In Section 4 the procedure of optimal stochastic UIO design, ensure the mean-square estimation error minimization is considered, moreover, the recurrent equations for non-stationary observer optimal parameters are founded. The conditions of steady-state stochastic UIO existence are established and optimal parameters of stationary stochastic UIO are founded based on the matrix Riccati equation solution. In order to ensure the above conditions fulfillment the UIO regularization approach was used.

Section 5 is devoted to solution of the problem of estimates fusion in networks of decentralized stochastic UIO. Two main information fusion schemes are considered: optimal estimates fusion in distributed UIO and estimates fusion in decentralized consensus UIO. In the first case the estimates aggregation procedure is realized, and in the second one, the averaging Consensus Protocol is applied and observer's parameter optimization procedure for the set of stochastic consensus UIO is based on the method of the state space expanding.

6.2 Decentralized State and Input Estimation Problem

Consider the problem of simultaneous decentralized state and input signal estimation for discrete-time stochastic dynamic system

$$
\begin{aligned}
x(k+1) &= Ax(k) + Bu(k) + w(k), \\
y(k) &= Cx(k) + v(k), \\
y_i(k) &= C_i x(k) + v_i(k), \quad i = \overline{1, N},
\end{aligned}
\tag{6.1}
$$

where $x(k)$ – state vector, $u(k)$ – input signal, $w(k)$ – system random disturbances, $y(k)$ – output measurement vector, divided into N individual measurement vectors $y_i(k)$, available in sensor network nodes, $v(k)$ – vector of random measurement errors (measurement noise), consist of N sub-vectors $v_i(k)$.

In the decentralized scheme for state/input estimation based on sensor and observers networks, each observer from the set of N agents produces its own local estimates on the basis of locally available measurements

$y_i(k)$, determined by the measurement matrix C_i with appropriate random measurement errors $v_i(k)$.

The system stochastic disturbances and random measurement errors are assumed to be a discrete white noise, i.e. for any sensor node:

$$\mathbf{E}\{w(k)\} = 0, \qquad \mathbf{E}\{w(k)w^{\mathrm{T}}(l)\} = Q\delta_{kl},$$
$$\mathbf{E}\{v_i(k)\} = 0, \qquad \mathbf{E}\{v_i(k)v_i^{\mathrm{T}}(l)\} = R_i\delta_{kl}, \qquad (6.2)$$
$$\mathbf{E}\{v_i(k)v_j^{\mathrm{T}}(l)\} = 0, \quad \mathbf{E}\{w(k)v_i^{\mathrm{T}}(l)\} = 0.$$

where Q, R_i – positive definite covariance matrices.

The problem of simultaneous state/input estimation consists in constructing real-time estimates $\hat{x}(k)$, $\hat{u}(k)$ through a set of observers or filters, using available measurements.

Classical approach to state vector estimation based on the theory of Luenberger observers and Kalman filters based on the assumption that the input signal $u(k)$ is precisely known. An essential feature of problem under consideration is the assumption that for system (6.1) input signals $u(k)$ is unknown and not directly measurable.

Problem of finding a state vector estimate $\hat{x}(k)$ under the unknown input signal will be called *Invariant State Estimation Problem*. The problem of constructing unknown input signal estimate $\hat{u}(k)$ can be interpreted as *System Inversion Problem*.

An effective way to solve this problem is to use the unknown-input observers (UIO) technique as well as inverse dynamical system theory.

Let that the following conditions take place: rank $B = m$, rank $C_i = p_i$, rank$(C_iB) = m$, $m \le p_i$, $i = \overline{1, N}$, which guarantee the existence the set of relevant UIO [23], [24].

Then the problem of decentralized state/input estimate for system (6.1) is reduced for the problem of set of UIO optimal design.

Consider a decentralized estimation problem for two main types of estimates fusion architecture:

a) Centralized estimation fusion scheme. In such a case the appropriate local state estimates are produced by a number of decentralized UIO observers, which use only locally available measurements:

$$\widetilde{x}_i(k+1) = F_i \cdot \widetilde{x}_i(k) + G_i \cdot y_i(k),$$
$$\hat{x}_i(k) = \widetilde{x}_i(k) + H_i \cdot y_i(k), \;\; i = \overline{1, N}, \tag{6.3}$$

or, in equivalent form,

$$\hat{x}_i(k+1) = F_i \cdot \hat{x}_i(k) + (G_i - F_i H_i) \cdot y_i(k) + H_i \cdot y_i(k+1). \tag{6.4}$$

The structure features of UIO is the existence of two information transmission channels, which provides the possibility of obtaining estimates that are invariant subject to the immeasurable input signal.

The resulting fusion aggregate state estimate formed in fusion center, is taken as a linear combination of partial decentralized state estimates

$$\hat{x}(k) = \sum_{i=1}^{N} \lambda_i \hat{x}_i(k), \;\; \sum_{i=1}^{N} \lambda_i = 1, \; \lambda_i > 0. \tag{6.5}$$

The problem of observers (6.5) optimal design is to find the set of matrices of observer's parameters $F_i, G_i, H_i, \; i = \overline{1, N}$, which ensure the "invariance properties" [25, 26] of observers (6.5), (6.6), i.e. state and input estimation error independence from the unknown and immeasurable input signal $u(k)$ and, moreover, minimizes the set of mean-square error state estimation performance indices

$$J_{x_i}(k+1) = \mathbf{E}\{e_{x_i}^T(k+)e_{x_i}(k+1)\},$$
$$e_{x_i}(k) = x_i(k) - \hat{x}_i(k), \; i = \overline{1, N}, \tag{6.6}$$
$$J_x(k+1) = \sum_{i=1}^{N} J_{x_i}(k+1).$$

For the final evaluation of aggregate optimal state estimate the aggregation weights $\lambda_i, \; i = \overline{1, N}$ should be found as a solution of aggregate performance index $J_x(k+1)$ minimization problem taking into account conditions (6.5).

b) Decentralized consensus observer scheme. For this scheme a network of consensus observers is used [12, 16]:

$$\hat{x}_i(k+1) = F_i \cdot \hat{x}_i(k) + (G_i - F_iH_i) \cdot y_i(k) + H_i \cdot y_i(k) -$$

$$- K_i \cdot [\hat{x}_i(k) - \frac{1}{N_i}\sum_{j \in V_i}\hat{x}_j(k)], \quad N_i = |\mathbf{V}_i|, \qquad (6.7)$$

$$\hat{x}(k) = \frac{1}{N}\sum_{i=1}^{N}\hat{x}_i(k), \quad i = \overline{1, N},$$

where K_i, $i = \overline{1, N}$ – consensus terms tuning matrices of observer network, $\mathbf{V}_i$ – set of agent-observers, which can supply its local estimation information $\hat{x}_i(k)$ to i-th observer in accordance to the network communication topology, N_i – number of elements in the set of neighbor observers, N – total number of agent-observers.

The problem is to find observers matrix parameters F_i, G_i, H_i, $i = \overline{1, N}$, as well as consensus term gain matrices K_i, which ensure the aggregate performance index $J_x(k+1)$ minimization.

Thus, in both cases, the problem is reduced to the synthesis of a set of optimal stochastic unknown-input observers followed optimal aggregation of their local estimates. As will be shown below, input signals estimate can be found using obtained state estimates via the inverse model approach.

The optimal design method of unknown-input observer network with information fusion includes three steps:

- Observers structural design in accordance to the requirement of estimation error independence from state and unknown input signal;
- Observers parameters as well as consensus gain optimization using the set of performance indexes, which characterizes the estimation accuracy;
- Aggregate estimate optimization thorough the optimal partial estimates linear combination.

6.3 State and input signal invariant estimation

6.3.1 Reduced order regularized observer design

Consider state and input estimation problem in deterministic statement with reference to discrete time system without stochastic disturbance and measurement noise:

$$x(k+1) = Ax(k) + Bu(k),$$
$$y(k) = Cx(k),$$

(6.8)

where $x(k) \in \mathbf{R}^n$ – system state vector at time instant k, $u(k) \in \mathbf{R}^p$ – vector of unknown and immeasurable input signals, $y(k) \in \mathbf{R}^q$ – vector of output measured signals.

State and input estimation problem includes finding the estimates $\hat{x}(k)$, $\hat{u}(k-d)$ using the available measurements $\{y(k)\}$, where integer $d \geq 1$ determines the minimal input signal estimation delay and concise with system (6.8) relative order, namely, minimal integer, such as matrix Markov parameter $S(d) = CA^{d-1}B \neq 0$.

The proposed approach to simultaneous estimation of state and unknown input signal is closely related with the problem of dynamical inversion of system (6.8).

Discrete dynamic system

$$\bar{x}(k+1) = A^I \bar{x}(k) + B_1^I y(k) + B_2^I y(k+d),$$
$$\hat{u}(k) = C^I \bar{x}(k) + D_1^I y(k) + D_2^I y(k+d)$$

(6.9)

with state vector $\bar{x}(k) \in \mathbf{R}^{n-q}$ will be referred to as a *reduced order inverse model* of (6.8), if system (6.9) is asymptotically stable and the following conditions take place: $\|\bar{x}(k) - Rx(k)\| \to 0$, $\|\hat{w}(k) - w(k)\| \to 0$, $k \to \infty$, where $R \in \mathbf{R}^{n-q \times n}$ is the appropriate aggregate matrix, such that rank $\left(C^T : R^T \right) = n$, rank $R = n - q$.

For the purpose of simplicity, consider the only case of system (6.8) relative order 1, then $S(1) = CB \neq 0$.

At that inverse model (6.9) output vector $\hat{u}(k)$ may be treated as input signal estimate and an inverse system (6.9) actually becomes an observer of the unknown input signal. Inverse model (6.9) state vector $\bar{x}(k)$ may be used, as it will be shown later, for system (6.8) state estimation.

Without loss of generality we can assume, that the dynamic system (6.8) model input and output matrices are of full rank, namely, rank $B = p$, rank $C = q$, and for system (6.8) invertability conditions [23] $S = CB \neq 0$, rank $S = q$, $q \geq p$ take place.

If such a case the structure inversion algorithm may be applied [31], which enables the construction of inverse models with parameters strictly determined by the parameters of system (6.8). This raises a number of difficulties, for example, for non-minimum phase system the inverse models will be unstable.

The minimal state-space realization of inverse model with desired dynamic properties may be obtained by means of reduced order UIO and then the observer equation combined with the unknown input signal estimate may be treated as the designed inverse model.

From practical point of view it is desirable to decompose the inverse model design problem solution into two steps, namely, the inverse model *structural synthesis* renders the fixed and free inverse model matrix parameters, and inverse model *parametric design* which ensure the inverse model stability and desired dynamic properties using the appropriate parameters tuning methods, such as root-locus, pole-placement or performance optimization.

Let $z = Rx(k) \in \mathbf{R}^{n-q}$ be a vector of aggregated auxiliary variables, where matrix $R \in \mathbf{R}^{n-q \times n}$ is the appropriate aggregate matrix.

Treated $u(k)$ as an unknown input signal, the state vector estimate $\hat{x}_k \in \mathbf{R}^n$ may be obtained by minimal-order UIO as follows:

$$\tilde{x}(k) = \overline{F}\tilde{x}(k) + \overline{G}y(k),$$
$$\overline{x}(k) = \tilde{x}(k) + \overline{H}y(k), \qquad (6.10)$$
$$\hat{x}(k) = Py(k) + Q\overline{x}(k),$$

where $\tilde{x}(k) \in \mathbf{R}^{n-q}$ – observer state vector, and matrices $P_{n \times q}$ and $Q_{n \times n-q}$ are uniquely determined by selected aggregating matrix R and has the following properties:

$$\begin{pmatrix} P & Q \end{pmatrix} = \begin{pmatrix} C \\ R \end{pmatrix}^{-1}, \quad PC + QR = I_n, \qquad (6.11)$$

$$CP = I_q, \quad RQ = I_q, \quad CQ = 0_{q \times n-q}, \quad RP = 0_{n-q \times q}.$$

From the equation of state estimate error the observer design conditions, ensuring state estimation error independence from unknown input, may be obtained in the form of the system of linear matrix equations:

$$\begin{aligned} &\left(R - \overline{H}C\right)A - \overline{F}\left(R - \overline{H}C\right) - \overline{G}C = 0, \\ &\left(R - \overline{H}C\right)B = 0. \end{aligned} \tag{6.12}$$

The solution of linear matrix equation (6.12) under the invertability conditions fulfillment may be obtained as

$$\begin{aligned} \overline{F} &= R\Pi AQ, \\ \overline{G} &= R\Pi A\left(H + P\Omega\right), \\ \overline{H} &= RBS^{+} = RH, \end{aligned} \tag{6.13}$$

where projection matrices

$$\Pi = I_{n} - BS^{+}C, \quad \Omega = I_{q} - SS^{+}, \quad C\Pi = \Omega C, \tag{6.14}$$

and "+" denotes Moore-Penrouze generalized matrix inverse [32].

Taking an unknown input signal estimate from (6.8), (6.10) as $\hat{u}(k) = B^{+}\left(\hat{x}(k+1) - A\hat{x}(k)\right)$, the reduced-order inverse model equations may be presented in the following form:

$$\begin{aligned} \overline{x}(k+1) &= \overline{F} \cdot \overline{x}(k) + R\Pi AP \cdot y(k) + \overline{H} \cdot y(k+1), \\ \hat{u}(k) &= B^{+}(H + P\Omega) \cdot [y(k+1) - CAQ \cdot \overline{x}(k) - CAP \cdot y(k)]. \end{aligned} \tag{6.15}$$

Thus the obtained inverse model (6.15) forms the state and input signal estimates $\hat{x}(k) = Py(k) + Q\overline{x}(k)$, $\hat{u}(k)$ and may be considered as invariant state and input observer. Indeed, the deviation vector $\overline{e}_{x}(k) = Rx(k) - \overline{x}(k)$ and unknown input signal estimation error $\overline{e}_{u}(k) = u(k) - \overline{u}(k)$ will be invariant with respect to unknown input

$$\begin{aligned} \overline{e}_{x}(k+1) &= \overline{F} \cdot \overline{e}_{x}(k), \\ \overline{e}_{u}(k) &= B^{+}\left(Q\overline{F} - AQ\right) \cdot \overline{e}_{x}(k), \end{aligned} \tag{6.16}$$

that grounds the proposed state and input estimation problem solution.

The inverse model dynamics matrix $A^I = \overline{F} = R\Pi AQ$ depends from the arbitrary aggregating matrix R of given rank equal to $n - q$, which may be used as a tuning matrix.

Using the special form of system (6.8) matrices

$$A = \begin{pmatrix} A_{11} & A_{12} \\ A_{21} & A_{22} \end{pmatrix}, \quad C = \begin{pmatrix} I_q & 0_{n-q\times q} \end{pmatrix}, \quad B = \begin{pmatrix} B_1 \\ B_2 \end{pmatrix}_{n-q}^{q}, \qquad (6.17)$$

which may be obtained by nonsingular state-space transformation, and concretely define the matrices P, Q choice, one can admit

$$\begin{pmatrix} P & Q \end{pmatrix} = \begin{pmatrix} P_1 & Q_1 \\ P_2 & Q_2 \end{pmatrix}_{n-q}^{q}, \quad P_1 = I_q, \quad Q_1 = 0_{q\times n-q}. \qquad (6.18)$$

In such a case, for any Q_2 such that $\det Q_2 \neq 0$, aggregating matrix may be found in the form $R = Q_2^{-1}\begin{pmatrix} -P_2 & I_{n-q} \end{pmatrix}$, and consequently the UIO matrices for system (6.8) representation (6.17) are the following:

$$\overline{F} = R\Pi AQ = Q_2^{-1}\begin{pmatrix} \tilde{A}_{22} - P_2\Omega_{B_1}A_{12} \end{pmatrix}Q_2,$$

$$\tilde{A}_{22} = A_{22} - B_2 B_1^+ A_{12}, \qquad (6.19)$$

$$\tilde{A}_{12} = \Omega_{B_1}A_{12}, \quad \Omega_{B_1} = I_q - B_1 B_1^+.$$

Thus, in fact, the nonsingular matrix Q_2 specifies the similarity transformation and doesn't change the A^I spectrum, which, as it follows from (6.12), completely determined by only arbitrary tuning matrix P_2.

Therefore, the aggregate matrix R is determined up to an arbitrary nonsingular matrix Q_2. The tuning matrix P_2 may be choosing by any type of pole placement method, and the problem will be solvable if matrix pair $[\tilde{A}_{22}, \tilde{A}_{12}]$ is observable. It may be shown, that such a condition is equivalent to the UIO design solvability condition, namely input observability [23, 24].

The UIO structural design solvability condition is obviously violated in the case, when $q = p$. At that $\Omega_{B_1} = 0$, and observer dynamic matrix $\overline{F}$ doesn't depend from tuning matrix P_2, which eliminate the possibility of its dynamic properties changing.

In such a case, to ensure tuning properties feasibility, it is expediently to use so-called "regularized" UIO [25, 26], which provide the approximate observer invariance with respect to unknown input signal and ensure fulfillment of the design solvability conditions.

To implement the regularization properties, an exact condition of invariance (6.12) is replaced by approximate one:

$$\|RB - HCB\|^2 + \varepsilon\|H\|^2 \to \min_{H}, \tag{6.20}$$

where $\varepsilon > 0$ – regularization parameter.

The regularized observer matrix parameters are determined from quadratic optimization problem (6.20) solution and structural design matrix equations (6.12).

Using (6.15), (6.20) it is easy to get the regularized solution of inverse model structural synthesis problem as follows:

$$\begin{aligned}
\overline{F}(\varepsilon) &= R\Pi(\varepsilon)AQ, \\
\overline{G}(\varepsilon) &= R\Pi(\varepsilon)A\big(\overline{H}(\varepsilon) + P\Omega(\varepsilon)\big), \\
\overline{H}(\varepsilon) &= RBS^+(\varepsilon),
\end{aligned} \tag{6.21}$$

where $\Pi(\varepsilon) = I_n - H(\varepsilon)C$, $\Omega_{B_1}(\varepsilon) = I_q - B_1 B_1^+(\varepsilon)$, and consequently the regularized projection matrices $\Pi(\varepsilon)$ and $\Omega_{B_1}(\varepsilon)$ has the following form:

$$\Pi(\varepsilon) = \begin{pmatrix} \Omega_{B_1}(\varepsilon) & 0_{q,n-q} \\ -B_2 B_1^+(\varepsilon) & I_{n-q} \end{pmatrix}, \tag{6.22}$$

$$\Omega_{B_1}(\varepsilon) = I_q - B_1 B_1^+(\varepsilon), \quad B_1^+(\varepsilon) = B_1^{\mathrm{T}}\big(\varepsilon I_q + B_1 B_1^{\mathrm{T}}\big)^{-1}.$$

Matrix $S^+(\varepsilon) = S^{\mathrm{T}}\big(\varepsilon I_q + SS^{\mathrm{T}}\big)^{-1}$ may be considered as a regularized inverse of matrix S, at that $S^+(\varepsilon)\big|_{\varepsilon=0} = S^{-1}$ and $S^+(\varepsilon)\big|_{\varepsilon=\infty} = 0$.

As a result, dynamics matrix $\overline{F}(\varepsilon)$ of regularized reduced-order inverse model defines as:

$$
\begin{aligned}
\overline{F}(\varepsilon) &= \tilde{A}_{22}(\varepsilon) - P_2\tilde{A}_{12}(\varepsilon), \\
\tilde{A}_{22}(\varepsilon) &= A_{22} - B_2 B_1^+(\varepsilon)A_{12}, \\
\tilde{A}_{12}(\varepsilon) &= \Omega_{B_1}(\varepsilon)A_{12}.
\end{aligned}
\tag{6.23}
$$

Finally, taking into account, that $\Omega_{B_1}(\varepsilon) = \varepsilon\left(\varepsilon I_q + B_1 B_1^{\mathrm{T}}\right)^{-1}$, from (6.21), (6.22) follows reduced-order regularized inverse model equations:

$$
\begin{aligned}
\overline{x}(k+1) &= \overline{F}(\varepsilon)\cdot\overline{x}(k) + \overline{L}(\varepsilon)\cdot y(k) + \overline{H}(\varepsilon)\cdot y(k+1), \\
\hat{u}(k) &= \overline{B}(\varepsilon)\cdot\left[y(k+1) - CAQ\cdot\overline{x}(k) - CAP\cdot y(k)\right],
\end{aligned}
\tag{6.24}
$$

where the regularized inverse model parameter matrices are

$$
\begin{aligned}
\overline{L}(\varepsilon) &= R\Pi(\varepsilon)AP, \quad \overline{H}(\varepsilon) = RBS^+(\varepsilon), \\
\overline{B}(\varepsilon) &= B^+\left(H(\varepsilon) + P\Omega(\varepsilon)\right), \quad \Omega(\varepsilon) = I_q - SS^+(\varepsilon).
\end{aligned}
\tag{6.25}
$$

Since the following directly verifiable matrix equalities take place

$$
\Pi(\varepsilon)B = B\left(I_n - S^+(\varepsilon)S\right) = \varepsilon B\left(\varepsilon I_n + S^{\mathrm{T}}S\right)^{-1},
\tag{6.26}
$$

the state and input estimation error satisfy the following difference equations:

$$
\begin{aligned}
\overline{e}_x(k+1) &= \overline{F}(\varepsilon)\cdot e(k) + \varepsilon RB\left(\varepsilon I_n + S^{\mathrm{T}}S\right)^{-1}\cdot u_k(k), \\
\overline{e}_u(k) &= B^+\left(QF(\varepsilon) - AQ\right)\cdot\overline{e}_x(k) + \varepsilon B^+QRB\left(\varepsilon I_n + S^{\mathrm{T}}S\right)^{-1}\cdot u(k).
\end{aligned}
\tag{6.27}
$$

It is obvious, that under $q = m$, regularized projection matrix $\Omega_{B_1}(\varepsilon) \neq 0$ for any $\varepsilon > 0$ and inverse model design problem become solvable, if pair matrix $[\tilde{A}_{22}(\varepsilon),\ \tilde{A}_{11}(\varepsilon)]$ is observable.

Solution of the problem of observer parametric design can be obtained on the basis of known modal control methods subject to the above conditions observability, which may be regarded as the conditions of UIO and inverse model parametric design problem solvability.

The regularization parameter ε should be selected based on trade-off between the desired observer dynamic properties, degree of stability in particular, and value of additional dynamic error component, proportional to the input signal, caused by inverse model regularization.

The advantage of considered method is in the fact that the designed inverse model has the desired dynamic properties and may be stabilized even for non-minimum phase system (6.8). Such an approach also may be used for UIO and inverse model stochastic optimization as well as state and input signals optimal estimation in the presence of random noise.

6.3.2   Input signal identification

The considered method for state and input estimation based on designed inverse model ensures the possibility of input signal real time estimation without necessity of essential *a priory* information about signal model. If some information concern input signal is available, the identification problem may be stated and solved within consider framework, since the current input signal estimates at any time instant may be treated as its *on-line* indirect measurements, perhaps with some measurement error.

For the identification purposes the input signal model should be previously selected. For the identification of complex shape signals it is necessary to use relatively complex models that reflect their characteristics features. Recently, as such a signal models are often used so-called wave models, describing non-periodic oscillation processes that frequently occur in practical applications.

Consider the input signal model (a) as a superposition of number of single harmonics with arbitrary unknown amplitudes and frequencies and corresponding indirect input signal measurement model (b):

$$(a)\quad u(k) = \sum_{j=0}^{m-1} u_j(k) = \sum_{j=0}^{m-1} \left[a_j \cos(\omega_j k) + b_j (\sin \omega_j k) \right]$$

$$(b)\quad \hat{u}(k) = u(k) + \zeta(k),$$

(6.28)

where $u_j(k)$ – scalar harmonic component of input signal $u(k)$ with frequencies $\{\omega_i\}$, $i = \overline{1, m}$, $0 < \omega_j = 2\pi f_j T_0 < \pi$, and amplitudes $\{a_i\}, \{b_i\}$, T_0 – sampling period, m – number of model harmonics, $\hat{u}(k)$ – input signal estimate, $\zeta(k)$ – stochastic variable, modeling input signal indirect measurement error, k – time instant.

In details the structure and properties of input signal indirect measurement errors will be analyze below in considering the problem of stochastic observer design.

In such a case, the identification problem is to find the estimates of the unknown parameters $\{a_i\}, \{b_i\}, \{\omega_i\}$, $i = \overline{1, m}$ of the input signal model.

Treating the input signal model harmonic component as a solution of the second order difference equation and using z-transformation, the input signal indirect measurement model (6.28) may be represented as:

$$\prod_{j=0}^{m-1}\left[1 - 2\cos(\omega_j z^{-1}) + z^{-2}\right] \cdot \hat{u}(k) = \zeta(k). \tag{6.29}$$

Realizing the inverse transformation in time domain, the equation (6.29) may be represented in the linear auto-regression form, which may be considered as an equivalent dynamic model of input signal (6.28).

Then the equivalent dynamic multi auto-regression input signal measurement equation takes the form:

$$\hat{u}(k) = \sum_{j=0}^{m-1}\beta_j\left(\hat{u}(k + j - m) + \hat{u}(k - j - m)\right) - \hat{u}(k - 2m) + \zeta(k) = \tag{6.30}$$

$$= \beta^T U(k,m) - \hat{u}(k - 2m) + \zeta(k),$$

where

$$U(k,m) = [2\hat{u}(k - m), \hat{u}(k - m + 1) + \hat{u}(k - m - 1), \ldots, \hat{u}(k - 1) + \hat{u}(k - 2m + 1)]^T$$

is the signal "prehistory" vector, $\beta^T = (\beta_0, \beta_1, \ldots, \beta_{m-1})$ – input signal equivalent auto-regression model parameters.

To solve the above identification problem the well-known recurrent identification algorithms may be used, which ensure *on-line* input signal model (6.28) parameters estimation.

Introduce quadratic identification criterion

$$J_u = \sum_{k=2m}^{N-1}\left[\hat{u}(k) + \hat{u}(k-2m) - \beta^{\mathrm{T}}U(k,m)\right]^2, \qquad (6.31)$$

the corresponding recurrent algorithm for input signal identification, which minimize (6.31), may be obtained:

$$\hat{\beta}(k) = \hat{\beta}(k-1) + [\hat{u}(k) + \hat{u}(k-m) -$$
$$- \hat{\beta}^{\mathrm{T}}(k-1)U(k,m)] \cdot u(k-m) \cdot r^{-1}(k), \qquad (6.32)$$
$$r(k) = \gamma(k)r(k) + \|U(k,m)\|^2, \quad 0 < \gamma(k) < 1,$$

where tuning parameter $\gamma(k)$ determining the step factor $r(k)$ of algorithm, should be selected based on the trade-off between tracking and flittering properties of the identification algorithm (6.32).

Frequencies estimates of signal model harmonics ω_j may be calculated by parameters β_j estimates using the following equation:

$$\beta_0 + \sum_{j=1}^{m-1}\beta_j\cos(j\omega) = \cos(m\omega). \qquad (6.33)$$

Taking into account the trigonometric identity
$$\cos m\omega = \cos^m \omega - C_m^2 \cos^{m-2}\omega \cdot \sin^2 \omega + C_m^4 \cos^{m-4}\omega \cdot \sin^4 \omega + ..., \qquad \text{the}$$
input signal model frequencies ω_j may be determined as a roots of power polynomial from the argument $\cos\omega$.

Similarly at any time instant k the estimates of wave model component harmonics amplitudes $\Theta = (a_0, a_1, ..., a_{m-1}, b_0, b_1, ..., b_{m-1})^{\mathrm{T}}$ may be obtained by the quadratic identification criterion minimization

$$J_\Theta = \|\overline{U}(k,m) - \Phi(k,m)\Theta\|^2 \qquad (6.34)$$

where the extended vector $\overline{U}(k,m)$ and matrix $\Phi(k,m)$ are defined as:

$$\overline{U}(k,m) = \begin{bmatrix} \hat{u}(2m) \\ \hat{u}(2m+1) \\ \vdots \\ \hat{u}(k) \end{bmatrix},$$

$$\Phi(k,m) = \begin{bmatrix} \cos 2m\hat{\omega}_0 & \cdots & \cos k\hat{\omega}_0 \\ \vdots & \cdots & \vdots \\ \cos 2m\hat{\omega}_{m-1} & \cdots & \cos k\hat{\omega}_{m-1} \\ \sin 2m\hat{\omega}_0 & \cdots & \sin k\hat{\omega}_0 \\ \vdots & \cdots & \vdots \\ \sin 2m\hat{\omega}_{m-1} & \cdots & \sin k\hat{\omega}_{m-1} \end{bmatrix}$$

and $\hat{\omega}_j, \ j = \overline{0, m-1}$ are the frequencies estimates, obtained from (6.33), where parameters $\beta^{\mathrm{T}} = (\beta_0, \beta_1, ..., \beta_{m-1})$ are replaced by its estimates $\hat{\beta}^{\mathrm{T}} = (\hat{\beta}_0, \hat{\beta}_1, ..., \hat{\beta}_{m-1})$ obtained by identification algorithm (6.32).

As a result in accordance with general Least Squares Method the input signal model parameters estimates defined as

$$\hat{\Theta}(k) = \left(\Phi(k,m)\Phi(k,m)^{\mathrm{T}} \right)^{-1} \Phi(k,m)U(k,m). \tag{6.35}$$

Formula (6.35) gives, in fact, the solution of input signal identification problem based on its decomposition into the problem of the input signal indirect measurement via inverse model with following signal model parameters estimation.

6.4 Stochastic optimal unknown input observer design

6.4.1 Observer structural design

Consider the problem of structural design for full-order UIO for system (6.1), (6.2):

$$\tilde{x}(k+1) = F\tilde{x}(k) + Gy(k),$$
$$\hat{x}(k) = \tilde{x}(k) + Hy(k). \tag{6.36}$$

From the equation for state estimation error $e_x(k) = x(k) - \hat{x}(k)$, which follow from system (6.1) and observer (6.36) equations, the "invariance condition" may be obtained, which ensures the independence of state estimation error from both state and unknown input.

Such a condition has the form of linear matrix equations:

$$(I_n - HC)A - F(I_n - HC) + GC,$$
$$B = HCB. \tag{6.37}$$

The solution of linear matrix equation (6.37) gives the observer parameters matrices, defined up to any arbitrary matrix, which may be treated as a matrix of observer tuning parameters

$$F = \Pi A - G_0 C, \quad G = \Pi AH + G_0 \Omega,$$
$$H = B(CB)^+, \quad \Pi = I_n - HC, \quad \Omega = I_p - CH, \tag{6.38}$$

where "+" – Moore-Penrose generalized matrix inversion, and G_0 is a corresponding tuning matrix.

In such a case the estimation error equation for observer (6.6) takes the following form

$$e_x(k+1) = F \cdot e_x(k) + \xi(k+1),$$
$$\xi(k+1) = \Pi \cdot w(k) - G_0 \cdot v(k) - H \cdot v(k+1), \tag{6.39}$$

where $\xi(k)$ is an equivalent random disturbance.

6.4.2   Stochastic observer optimization

To solve the problem of optimal stochastic UI-observer design preliminary obtain the expression for the state estimate performance index as an estimate mean-square error.

Using the covariance matrix equation for state estimate

$$P_e(k+1) = FP_e(k)F^{\mathrm{T}} + FHRG_0^{\mathrm{T}} +$$
$$+ G_0 RHF^{\mathrm{T}} + G_0 RG_0^{\mathrm{T}} + HRH^{\mathrm{T}} + \Pi Q \Pi^{\mathrm{T}}, \tag{6.40}$$

it is possible to determine the estimation accuracy as an state estimate mean square error

$$J_x(k+1) = \mathrm{tr}\{G_0 W_0(k)G_0^{\mathrm{T}} - G_0 W_1^{\mathrm{T}}(k) - W_1(k)G_0^{\mathrm{T}} + W_2(k)\}, \tag{6.41}$$

where matrices $W_0(k)$, $W_1(k)$, $W_2(k)$ depend on the covariance matrix of state estimation errors

$$W_0(k) = CP_e(k)C^{\mathrm{T}} - CHR - RH^{\mathrm{T}}C^{\mathrm{T}} + R,$$
$$W_1(k) = \Pi A[P_e(k)C^{\mathrm{T}} - HR], \tag{6.42}$$
$$W_2(k) = \Pi A P_e(k) A^{\mathrm{T}} \Pi^{\mathrm{T}} + HRH^{\mathrm{T}} + \Pi Q \Pi^{\mathrm{T}}.$$

The optimal observer tuning matrix $G_0^*(k)$ may be found from following optimization problem

$$G_0^*(k) = \arg\min J_x(k+1),$$
$$J_x(k+1) = \mathrm{tr}P_e(k+1). \tag{6.43}$$

Solution to the optimization problem (6.43) can be found directly and has the form $G_0^*(k) = W_1(k)W_0^+(k)$.

As a result, the optimal stochastic UIO equation takes the form

$$\hat{x}(k+1) = F^*(k) \cdot \hat{x}(k) + G_0^*(k) \cdot y(k) + B(CB)^+ \cdot y(k+1),$$
$$F^* = \Pi A - G_0^*(k)C, \quad G_0^*(k) = W_1(k)W_0^+(k). \tag{6.44}$$

Wherein the state estimate covariance matrix for stochastic UIO described by the matrix recurrent equation:

$$P_e(k+1) = \Pi A P_e(k) A^{\mathrm{T}} \Pi^{\mathrm{T}} - \Pi A[P_e(k)C^{\mathrm{T}} - HR] \cdot$$
$$\cdot W_0^+(k)[CP_e(k) - RH^{\mathrm{T}}]A^{\mathrm{T}} \Pi^{\mathrm{T}} + HRH^{\mathrm{T}} + \Pi Q \Pi^{\mathrm{T}}. \tag{6.45}$$

6.4.3   Optimal input signal estimate

To obtain an estimate of the input signal $u(k)$ use the state estimates $\hat{x}(k)$, generated by the observer (6.36). Substituting them into the system (6.1) equation, we obtain the dynamic regression equation for the input signal, which may be used for its optimal estimation. At that an equivalent model of indirect input signal measurement becomes

$$Bu(k) = \hat{x}(k+1) - A\hat{x}(k) + \eta(k+1),$$
$$\eta(k+1) = (F - A) \cdot e_x(k) - \xi(k+1) - w(k), \tag{6.46}$$

where equivalent input estimation disturbance $\eta(k)$ may be considered as the indirect measurement error.

An appropriate input signal estimate can be determined by the maximum likelihood scheme or least squares method. Since the covariance matrix $P_\eta(k)$ of equivalent input estimation disturbance $\eta(k)$ may be singular, in order to find the optimal LGQ input signal estimate it is necessary to use generalized Gauss-Markov estimate [32]:

$$\hat{u}(k) = B^*(k) \cdot [\hat{x}(k+1) - A\hat{x}(k)],$$
$$B^*(k) = B^+ \{I_m - P_\eta^{1/2}(k)[\Omega_B P_\eta^{1/2}(k)]^+\}, \quad \Omega_B = I_n - BB^+. \tag{6.47}$$

Equations (6.36), (6.47) together define the structure of optimal stochastic state and input observer and may be considered as an optimal stochastic inverse model of system (6.1):

$$\hat{x}(k+1) = F^* \cdot \hat{x}(k) + G_0^* \cdot y(k) + H^* \cdot y(k+1),$$
$$\hat{u}(k) = B^*(F - A) \cdot \hat{x}(k) + B^* \cdot G_0^* y(k) - (CB)^+ \cdot y(k+1). \tag{6.48}$$

Note that input signal estimates, generated by a designed optimal inverse model (6.48), are delayed by step regarding current measurements.

6.4.4 Steady-state optimal stochastic observer

For practical applications, it is desirable to use a steady-state unknown-input observer, the parameters of which are determined in turn by the steady-state covariance matrix of state estimation errors.

In order to obtain steady-state stochastic UIO existence conditions use the following estimation error covariance matrix transformation

$$P_e(k) = P(k) + \Psi, \quad \Psi C^\mathrm{T} = HR. \quad (6.49)$$

Then the equivalent Riccati equation becomes

$$P(k+1) = \Pi A P(k) A^\mathrm{T} \Pi^\mathrm{T} - \Pi A P(k) C^\mathrm{T} \cdot$$
$$\cdot [C^\mathrm{T} P(k) C + \overline{R}]^+ C P(k) A^\mathrm{T} \Pi^\mathrm{T} + \overline{Q}, \quad (6.50)$$

where transformed covariance matrices are

$$\overline{R} = \tfrac{1}{2}(\Omega R + R\Omega^\mathrm{T}),$$
$$\overline{Q} = \Pi A \Psi A^\mathrm{T} \Pi^\mathrm{T} - \Psi + HRH^\mathrm{T} + \Pi Q \Pi^\mathrm{T}. \quad (6.51)$$

Using the method of steady-state stochastic optimal filters analysis [33], the conditions of steady-state optimal stochastic UIO existence may be obtained in the following form [28]:

- There exist non-negative definite symmetric matrix which satisfies the equation $\Psi C^\mathrm{T} = HR$;
- Pair matrix $(\Pi A, C)$ is observable, i.e.

  $\mathrm{rank}[C^\mathrm{T} \quad (A^\mathrm{T}\Pi^\mathrm{T})C^\mathrm{T} \quad ... \quad (A^\mathrm{T}\Pi^\mathrm{T})^{n-1}C^\mathrm{T}] = n;$

- Pair matrix $(\Pi A, \overline{Q})$ is controllable, i.e.

  $\mathrm{rank}\left[\overline{Q}^{1/2} \quad (\Pi A)\overline{Q}^{1/2} \quad ... \quad (\Pi A)^{n-1}\overline{Q}^{1/2}\right] = n.$

Then there exist a unique positive definite solution $\overline{P}$ of linear matrix equation

$$\overline{P} = \Pi A \overline{P} A^\mathrm{T} \Pi^\mathrm{T} - \Pi A \overline{P} C^\mathrm{T} [C^\mathrm{T} \overline{P} C + \overline{R}]^{-1} C \overline{P} A^\mathrm{T} \Pi^\mathrm{T} + \overline{Q},$$
$$\lim_{k\to\infty} P(k) = \overline{P}, \quad \mathrm{tr}\,\overline{P} < \infty, \quad \overline{P}_e = \lim_{k\to\infty} P_e(k), \quad \mathrm{tr}\,\overline{P}_e < \infty. \quad (6.52)$$

The given condition of steady-state stochastic UI-observer existence fulfillment is determined by the properties of the projection matrix Π. It is easy to verify that in the case when $p = m$ the condition above doesn't fulfill. In such a case in order to ensure the steady-state observers existence the regularization approach for UIO design [25], [26] may be used. In accordance with such an approach the "invariance condition" (6.10) are replaced by approximate "sub-invariance" one

$$B - HCB = \varepsilon B, \quad 0 < \varepsilon < 1, \tag{6.53}$$

where ε is a regularization parameter.

Thus the appropriate solution of observer structural design problem takes the following form

$$
\begin{aligned}
F_\varepsilon &= \Pi_\varepsilon A - G_0(\varepsilon) C, \quad H_\varepsilon = (1 - \varepsilon) B (CB)^+, \\
G_\varepsilon &= \Pi_\varepsilon A H_\varepsilon + G_0(\varepsilon) \Omega_\varepsilon, \\
\Pi_\varepsilon &= I_n - H_\varepsilon C, \quad \Omega_\varepsilon = I_p - C H_\varepsilon.
\end{aligned} \tag{6.54}
$$

Following the above procedure we find the optimal values of the steady-state observer tuning parameters matrix by minimizing the estimation mean square error. For this purpose introduce *a priori* information about the unknown input signal in statistical form

$$\mathbf{E}\{u(k)\} = 0, \quad \mathbf{E}\{u(i)u^{\mathrm{T}}(j)\} = S_0 \delta_{ij}. \tag{6.55}$$

Then the optimal observer matrices obtained by steady-state performance index (6.41) optimization are given by

$$
\begin{aligned}
G_0^*(\varepsilon) &= \Pi_\varepsilon A \overline{P}_\varepsilon C^{\mathrm{T}} (C \overline{P}_\varepsilon C^{\mathrm{T}} + \overline{R}_\varepsilon)^{-1}, \\
B_0^*(\varepsilon) &= [B_\varepsilon^{\mathrm{T}} S_0^*(\varepsilon) B_\varepsilon]^{-1} B_\varepsilon^{\mathrm{T}} S_0^*(\varepsilon), \\
S_0^*(\varepsilon) &= (\varepsilon^2 I_n + \overline{S}_\varepsilon)^{-1}, \quad B_\varepsilon = (1 - \varepsilon) B,
\end{aligned} \tag{6.56}
$$

where matrix $\overline{P}_\varepsilon$ is a positive definite solution of the regularized stationary Riccati equation

$$\overline{P}_\varepsilon = \Pi_\varepsilon A \overline{P}_\varepsilon A^{\mathrm{T}} \Pi_\varepsilon^{\mathrm{T}} - \Pi_\varepsilon A \overline{P}_\varepsilon C^{\mathrm{T}} \cdot [C^{\mathrm{T}} \overline{P}_\varepsilon C + \overline{R}_\varepsilon]^{-1} C \overline{P}_\varepsilon A^{\mathrm{T}} \Pi_\varepsilon^{\mathrm{T}} + \overline{Q}_\varepsilon,$$

$$\overline{R}_\varepsilon = \tfrac{1}{2}[(I_p - CH_\varepsilon)R + R(I_p - CH_\varepsilon)^{\mathrm{T}}],$$

$$\overline{Q}_\varepsilon = \Pi_\varepsilon A \Psi_\varepsilon A^{\mathrm{T}} \Pi_\varepsilon^{\mathrm{T}} - \Psi_\varepsilon + H_\varepsilon R H^{\mathrm{T}} + \Pi_\varepsilon Q \Pi_\varepsilon^{\mathrm{T}} + \varepsilon^2 B S_0 B^{\mathrm{T}},$$

$$\Psi_\varepsilon C^{\mathrm{T}} = H_\varepsilon R,$$

$$(6.57)$$

and the steady-state covariance matrix $\overline{S}_\varepsilon$ is determined by the expression

$$\overline{S}_\varepsilon = \Pi_\varepsilon \Phi_1(\varepsilon) + \Phi_1(\varepsilon)\Pi_\varepsilon - \Pi_\varepsilon \Phi_2(\varepsilon)\Pi_\varepsilon^{\mathrm{T}} +$$

$$+ H_\varepsilon \Phi_2(\varepsilon)H_\varepsilon^{\mathrm{T}} + \varepsilon^2 B S_0 B^{\mathrm{T}},$$

$$\Phi_1(\varepsilon) = A \overline{P}_\varepsilon C^{\mathrm{T}} (C \overline{P}_\varepsilon C + \overline{R}_\varepsilon)^{-1} C \overline{P}_\varepsilon A^{\mathrm{T}},$$

$$\Phi_2(\varepsilon) = C A \overline{P}_\varepsilon A^{\mathrm{T}} C^{\mathrm{T}} + C A \Psi_\varepsilon A^{\mathrm{T}} C^{\mathrm{T}} + C Q C^{\mathrm{T}} + R.$$

$$(6.58)$$

Finally, the representation of regularized optimal stochastic UIO based inverse model takes the following form:

$$\hat{x}(k+1) = F_\varepsilon \cdot \hat{x}(k) + G_0^*(\varepsilon) \cdot y(k) + H_\varepsilon \cdot y(k+1),$$

$$\hat{u}(k) = B_0^*(\varepsilon)[(F_\varepsilon - A) \cdot \hat{x}(k) + G_0^*(\varepsilon) \cdot y(k) + H_\varepsilon \cdot y(k+1)].$$

$$(6.59)$$

The estimation accuracy performance index of input signal estimate as a mean-square error is given by

$$J_u(\varepsilon) = \mathrm{tr}\,\overline{P}_u(\varepsilon) = \mathrm{tr}\,B_0^*(\varepsilon)\overline{S}_\varepsilon[B_0^*(\varepsilon)]^{\mathrm{T}}, \qquad (6.60)$$

where $B_0^*(\varepsilon)$ is determined by.

The obtained expression (6.60) may be used to find the optimal value of the regularization parameter ε as the solution of function $J_u(\varepsilon)$ scalar optimization problem. Such a choice of the regularization parameter ε provides a compromise between the current information, obtained from available measurements and *a priori* information concern the estimated input signal, defined by its covariance matrix S_0.

6.5 Optimal information fusion in stochastic observer network

6.5.1 Information fusion for distributed stochastic observers

The problem of optimal decentralized state estimation aggregation in *Centralized Estimation Fusion Scheme* may be solved under assumption that the resulting aggregate estimate is a linear combination of partial estimates formed by local observers.

Suppose that the problem of decentralized state and input signal estimation for system (6.1) is solved by the set of independent UI-observers

$$\tilde{x}_i(k+1) = F_i \cdot \tilde{x}_i(k) + G_i \cdot y_i(k),$$
$$\hat{x}_i(k) = \tilde{x}_i(k) + H_i \cdot y_i(k), \quad i = \overline{1, N}. \quad . \tag{6.61}$$

If observers (6.61) parameters are selected from the following conditions of estimation errors $e_{i,x}(k) = x_i(k) - \hat{x}_i(k)$ independence from state and unknown and immeasurable input signal, namely, $F_i = \Pi_i A - G_{i,0} C_i$, $F_i = \Pi_i A - G_{i,0} C_i$, $G_i = \Pi_i A H_i + G_{i,0} \Omega_i$, $H_i = B(C_i B)^+$, the mentioned above estimation errors vectors will satisfy the system of difference equations

$$e_{i,x}(k+1) = F_i \cdot e_{i,x}(k) + \xi_i(k+1),$$
$$\xi_i(k+1) = \Pi_i \cdot w(k) - G_{i,0} \cdot v_i(k) - H_i \cdot v_i(k+1). \tag{6.62}$$

Let us choose the aggregated state vector estimate $\hat{x}(k)$ as linear combinations of partial estimates $\hat{x}_i(k)$, formed by local observers

$$\hat{x}(k) = \sum_{i=1}^{N} \lambda_i \hat{x}_i(k), \ \sum_{i=1}^{N} \lambda_i = 1, \ \lambda_i > 0, \tag{6.63}$$

where $\lambda = (\lambda_1 \ \lambda_2 \ ... \ \lambda_N)^{\mathrm{T}}$ are weighted coefficients.

Consider the aggregation performance index

$$J_\lambda(k) = \lambda^{\mathrm{T}} T(k) \lambda, \quad T(k) = [\mathrm{tr} \ P_{ij}(k)], \tag{6.64}$$

where corresponding auto and cross-covariance matrices of state estimation errors for different observers are determined by matrix difference equations

$$P_{ij}(k+1) = F_i P_{ij}(k) F_j^{\mathrm{T}} + \Xi_{ij}(k), \quad i, j = \overline{1, N},$$

$$P_{ij}(k) = \mathbf{E}\{e_{i,x}(k) e^{\mathrm{T}}_{i,x}(k)\}, \quad \Xi_{ij}(k) = \mathbf{E}\{\xi_i(k)\xi_j^{\mathrm{T}}(k)\}. \tag{6.65}$$

Note that matrix $\Xi_{ij}(k) = \Pi_i Q \Pi_j^{\mathrm{T}}$, so the estimation errors vectors $e_{i,x}(k)$, $e_{j,x}(k)$ are correlated one with each other. As a result, the optimal vector of aggregation weight parameters $\lambda = (\lambda_1 \, \lambda_2 \dots \lambda_N)^{\mathrm{T}}$ may be found as a solution of the following quadratic optimization problem

$$\lambda^{\mathrm{T}} T(k)\lambda \to \min, \quad \lambda^{\mathrm{T}} \mathbf{e} = 1, \quad \mathbf{e} = (1\,1\dots 1)^{\mathrm{T}}. \tag{6.66}$$

The solution of conditional optimization problem (6.66) may be obtained explicitly

$$\lambda^*(k) = (\mathbf{e}^{\mathrm{T}} T^{-1}(k)\mathbf{e})^{-1} T^{-1}(k)\mathbf{e}. \tag{6.67}$$

Finally the optimal aggregated state estimate obtained by decentralized stochastic UIO takes the following form:

$$\hat{x}(k) = (\mathbf{e}^{\mathrm{T}} T^{-1}(k)\mathbf{e})^{-1} \sum_{i=1}^{N} (\mathbf{e}_i^{\mathrm{T}} T^{-1}(k)\mathbf{e}) \cdot \hat{x}_i(k), \quad \mathbf{e}_i = (0 \dots 0\,1\,0 \dots 0)^{\mathrm{T}}. \tag{6.68}$$

At that the optimal unknown-input observers (61) tuning parameters $G^*_{i,0}$ are found, as before, by minimizing local performance indexes in the form of mean-square errors.

6.5.2　Information fusion in consensus observer network

For the *Decentralized Consensus Observer Scheme*, the problem of optimal state/input estimates fusion may be solved using a network of consensus unknown-input stochastic observer. At that the information exchange between the local observers can improve overall accuracy of estimation. Communication topology of sensor and observer network is described by finite graph $\mathbf{G}=(\mathbf{N},\mathbf{E})$, where $\mathbf{N}=\{i\}$, $|\mathbf{N}|=N$ denote a set of nodes (graph vertices), associated with measurements and data processing units (sensor-observer agents) and $\mathbf{E}=\{(i,j)\subseteq\mathbf{N}\times\mathbf{N}\}$ denote a set of edges, associated with data transmission lines.

Let $\mathbf{V}_i=\{j\in\mathbf{N}\,|\,(j,i)\in\mathbf{E}\}$ denote a set of neighbors of node i and $N_i=|\mathbf{V}_i|$, $i\in\mathbf{N}$ denote the degree (number of neighbors) for node i. Thus the observer located in node i may receive information (current estimates) from observers, located in nodes from the set $\mathbf{V}_i$.

The adjacency matrix $A=[\alpha_{ij}]\in\mathbf{R}^{N\times N}$ of graph $\mathbf{G}=(\mathbf{N},\mathbf{E})$, specified the interconnections in sensor and observer network, defines as

$$A=[\alpha_{ij}]\in\mathbf{R}^{N\times N}, \quad \alpha_{ij}=\begin{cases}1, & \text{if}\quad (j,i)\in\mathbf{E},\\ 0, & \text{otherwise.}\end{cases} \tag{6.69}$$

The Graf Laplasian matrix of $\mathbf{G}$ is defined as $L=\Delta(A)-A$, where $\Delta(A)$ is a degree matrix of $\mathbf{G}$ [34]:

$$\Delta(A)=\operatorname{diag}\begin{pmatrix}d_1 & d_2 & \dots & d_N\end{pmatrix}, \quad d_i=\sum_{j\in\mathbf{V}_i}\alpha_{ij}, \quad i,j\in\mathbf{N}. \tag{6.70}$$

Let decentralized state estimates are formed by a network of consensus observers:

$$\hat{x}_i(k+1)=F_i\cdot\hat{x}_i(k)+G_{i,0}\cdot y_i(k)+H_i\cdot y_i(k)-$$
$$-K\cdot[\hat{x}_i(k)-\frac{1}{N_i}\sum_{j\in\mathbf{V}_i}\hat{x}_j(k)], \quad i=\overline{1,N}, \tag{6.71}$$

where the observers parameters are chosen on the basis of UIO structural design problem solution by the described above procedure from the estimation errors invariance condition subject to immeasurable signals. In this case, $F_i=\Pi_i A-G_{i,0}C_i$, $H_i=B(C_iB)^+$, and observers tuning matrices $G_{i,0}$ as well as consensus gain matrix K are selectable. Observers' equations can be represented in equivalent form assuming, for problem simplification, that all the consensus gain matrices are equal.

$$\hat{x}_i(k+1) = F_i \cdot \hat{x}_i(k) + G_{i,0} \cdot y_i(k) + H_i \cdot y_i(k) -$$
$$- K \cdot \sum_{j \in V_i} [\hat{x}_i(k) - \hat{x}_j(k)], \quad i = \overline{1, N}. \tag{6.72}$$

The corresponding equations for estimation errors based on the UIO properties and subject to (39) are the following:

$$e_{i,x}(k+1) = F_i \cdot e_{i,x} + K \cdot \sum_{j \in V_i} [e_{i,x}(k) - e_{j,x}(k)] + \xi_i(k+1),$$
$$\xi_i(k+1) = \Pi_i \cdot w(k) - G_{i,0} \cdot v_i(k) - H_i \cdot v_i(k+1), \quad i = \overline{1, N}. \tag{6.73}$$

For any finite set of vectors $\{x_i(k)\}_{i=1}^{N}$, $x_i(k) \in \mathbf{R}^n$ denote $X(k) = (x_1^{\mathrm{T}}(k) \quad x_2^{\mathrm{T}}(k) \quad \ldots \quad x_N^{\mathrm{T}}(k))^{\mathrm{T}}$, and for any set of matrices $\{C_i\}_{i=1}^{N}, \in \mathbf{R}^{p \times n}$ denote $\tilde{C} = \sum_{i=1}^{N}(e_i e_i^{\mathrm{T}} \otimes C_i), \quad \tilde{C} \in \mathbf{R}^{pN \times nN}$, wherein $\tilde{C} = (I_N \otimes C), \tilde{C} \in \mathbf{R}^{n \times N}$ if $C_i = C$, where $\otimes$ – Kronecker product. Then the equation (6.73) can be written in expanded form:

$$E_x(k+1) = \tilde{F} \cdot E_x(k) + (L \otimes K)E_x(k) + \Xi(k),$$
$$= \Phi_{G,K} \cdot E_x(k) + \Xi(k), \tag{6.74}$$

Moreover, the dynamic matrix of the extended system $\Phi_{G,K} \in \mathbf{R}^{nN \times nN}$ depends on observers tuning parameters matrices $\{G_{1,0}, G_{2,0}, \ldots, G_{N,0}\}$ as well as consensus gain matrix K. From the equation (6.74) follows the equation for the covariance matrix $P_E(k) = \mathbf{E}\{E_x(k)E_x^{\mathrm{T}}(k)\}$ of the error estimation extended vector:

$$P_E(k+1) = \Phi_{G,K} \cdot P_E(k)\Phi_{G,K}^{\mathrm{T}} + P_\Xi,$$
$$P_\Xi(k) = \tilde{\Pi}Q_N\tilde{\Pi}^{\mathrm{T}} + \tilde{G}_0 R_N \tilde{G}_0^{\mathrm{T}} + \tilde{H}R_N \tilde{H}^{\mathrm{T}}, \tag{6.75}$$

where $Q_N = \text{block diag} \{Q\}$, $R_N = \text{block diag} \{R_i\}$, $R_N = \text{block diag} \{R_i\}$.

Finally the tuning matrix parameters of consensus UIO network (6.72) can be determined as a solution of the following optimization problem:

$$\{G_{1,0}^*, G_{2,0}^*, ..., G_{N,0}^*, K^*\} = \arg\min \bar{J}_{G,K} = \arg\min \operatorname{tr} \bar{P}_E, \quad (6.76)$$

where steady-state covariance matrix $\bar{P}_E$ is determined by the solution of Lyapunov matrix equation $\bar{P}_E = \Phi_{G,K} \cdot \bar{P}_E \Phi_{G,K}^T + P_\Xi$. If possible local state estimates aggregation, the resulting final estimate of the state vector can be taken as

$$\hat{x}(k) = \Lambda \hat{X}(k), \quad \Lambda = \begin{bmatrix} \lambda_1 I_n & \lambda_1 I_n & \cdots & \lambda_1 I_n \end{bmatrix}, \quad (6.77)$$

where optimal values of aggregation weight parameters $\lambda = (\lambda_1 \, \lambda_2 \, ... \, \lambda_N)^T$ is determined from the solution of the auxiliary optimization problem

$$\operatorname{tr} \Lambda \bar{P}_E \Lambda^T \to \min, \quad \lambda^T \mathbf{e} = 1, \quad \mathbf{e} = (1\,1\,...\,1)^T. \quad (6.78)$$

The designed UIO consensus network also provides the possibility of input signal estimate in accordance with the described above technique. Thus, the combination of the invariant and consensus estimation techniques offers the opportunities to efficient information fusion in multi-agent sensor-observer network, which in turn, improves the accuracy of state and input estimation.

References

[1] J. Manyika and H. Durrant-White. *Data Fusion and Sensor Management: a Decentralized Information-Theoretic Approach.* Ellis Horwood, 1994.

[2] V. Ligins, D. Hall and L. James. *Handbook of Multisensor Data Fusion. Theory and Practic.* New-York, Taylor & Fransis Group, 2007.

[3] I. Akyildis, W. Su, Y. Sankarasubramniam and E. Cayirci. A survey on sensor networks. *IEEE Communication Magazine*:102-114, August 2002.

[4] A. Makarenko, A. Brooks, S. Wiliams, H. Durrant-White and B. Grocholsky. *An Arhitecture for Decentralized Sensor Networks.* New Orleans, LA: IEEE ICRA, 2004.

[5] R. Luo, C.-C. Yih, and K. Su. Multisensor fusion and integration: approaches, applications, and future research directions. *IEEE Sensor Jornal*, 2(2):107-119, 2002.

[6] F. Castanedo. A review of data fusion techniques. *The Scientific World Jornal - Hindawi Publishing Corp.*, 2013.

[7] M. Ahmed. Decentralized state estimation in large-scale systems. *International Jornal of Systems Science*, 25(10):1577-1591, 1994.

[8] I. Kim, J. Chandrasekar, H. Palanthandalam-Madapusi and A. Redley. State estimation for large-scale systems based on reduced-order error-covariance propagation. In *Proceedings of the American Control Conference ACC'07*, 2007:5700-5705.

[9] M. Machmoud. Decentralized state-estimation of interconnected systems with unknown nonlinearities. *Jornal of Optimization Theory and Applications*, 152(3):786-798, 2012.

[10] R. Olfati-Saber, J. Fax and R. Murray. Consensus and cooperation in networked multi-agent systems. *Proceedings of the IEEE*, 95(1):215-233, September 2007.

[11] C. Yoshioka and T. Namerikawa. Observer-based consensus control strategy for multi-agent system with communication delay. In *Proceedings of 17th International Conference on Control Applications*, San Antonio, Texas, USA, 2008:1037-1042.

[12] D. Spanos, R. Olfati-Saber and R. Murray. Approximate distributed Kalman filter in sensor networks with quantifiable performance. In *Proceedings of 4-th International Symposium on Information Processing in Sensor Networks*, 2005:133-139.

[13] R. Olfati-Saber. Distributed Kalman filtering for sensor networks. In *Proceedings of 46-th IEEE Conference*, Dartmous Coll., Hanover, 2007:5492-5498.

[14] S. Sun and Z. Deng. Multi-sensor optimal information fusion Kalman filter. *Automatica*, 40:1017-1023, 2004.

[15] Z. Deng and R. Qi. Multi-sensor information fusion in suboptimal sready-state Kalman filter. *Chinese Science Abstracts*, 6(2):183-186, 200.

[16] B. Acikmese and M. Mandic. Decentralised observer with a consensus filter for distributed discrete-time linear systems. In *Proceedings of American Control Conference ACC'11*, 2011:4723-4730.

[17] R. Olfati-Saber and J. Shamma. Consensus filters for sensor networks and distributed sensor fusion. In *Proceedings of Joint Control and Decision Conference - European Control Conference CDC-ECC'05*, 2005:6698-6703.

[18] R. Olfati-Saber. Distributed Kalman filter with embedded consensus filter. In *Proceedings of the Joint Cointrol and Decision Conference - European Control Conference CDC-ECC'05*, 2005.

[19] L. Zhang and A. Cichocki. Blind deconvolution of dynamic systems: a stste space approach. *Jornal of Signal Processing*, 4(2):111-130, 2000.

[20] M. Hassan, M. Sultan and M. Attia. Fault detection in large-scale dymamic systems. *IEE Proceedings of Control Theory and Applications*, 139(2):119-124, 1992.

[21] J. Chang. Robust output feedback disturbance rejection control by simultaneously estimating state and disturbance. *Jornal of Control Science and Engineering*, 2011:13 p., 2011.

[22] H. Gao and L. Sun. Optimal disturbance rejection with zero steady-state error for time-delay large-scale systems with persistent disturbances. In *Proceedings of Control and Decision Conference CDC'09*, 2009:1922-1927.

[23] M. Hou and P. Mueller. Design of observers for linear systems with unknown inputs. *IEEE Transactions on Automatic Control*, 37:871-875, 1992.

[24] S. Zak. Observer design for systems with unknown inputs. *International Jornal Applied Mathematics and Computer Sciences*, 15(4), 2008.

[25] L. Lyubchyk and G. Grinberg. Blind deconvolution and separation signal processing via inverse model approach. In *Proc. of 4-th International Workshop on Intelligent Data Acquisition and Advanced Computer Systems IDAACS'2007*, Dortmund, Germany, 2007:325-328.

[26] L. Lyubchyk. Disturbance rejection in linear discrete multivariable systems: Inverse model approach. In *Proc. of 18-th IFAC World Congress*, Milan, Italy, 2011:7921-7926.

[27] M. Witczak and P. Pretci. Design of an extended unknown input observer with stochastic robustness techniques and evolutionary algorithms. *International Jornal of control*, 80(5):749-762, 2007.

[28] L. Lyubchyk. Optimal data fusion in decentralized stochastic unknown input oibservers. In *Proc. of 7-th International Workshop on Intelligent Data Acquisition and Advanced Computer Systems IDAACS'2013*, Berlin, Germany, 2013:358-362.

[29] S. Pilosu, A. Pisano and E. Usai. Decentralized state estimation for linear systems with unknown input: a consensus-based approach. *IET Control Theory Applications*, 5:498-506, 2001.

[30] S. Stankovic', M. Stankovic' and D. Stipanovic'. Consensus-based overlapping decentralized estimator. *IEEE Transaction on Automatic Control*, 54:410-415, 2009.

[31] H. Seraji. Minimal inversion, command mathing and disturbance rejection in multivariable systems. *IEEE Transactions on Automatic Control*, 14:270-276, 1989.

[32] A. Albert. *Regression and the Moore-Penrose Pseudoinverse*. New-York and London, Academic Press, 2003.

[33] R. Liptser and A. Shiryaev. *Statistics of Random Processes I. General Theory*. Springer-Verlag, 1978.

[34] C. Godsil and G. Royle. *Algebraic Graph Theory*. Springer, 2001.

Biography

Leonid Lyubchyk graduated from Kharkiv Polytechnic Institute, Ukraine, in 1973. From 1974 to 1976 he studied as a Ph.D. student in Institute of Control Problems, Moscow. In 1980 he joined the National Technical University "Kharkiv Polytechnic Institute", where in 1995 he obtained the Dr. Sc. degree. Since 2002 he holds the Professor position, and currently he is Head of Computer Mathematics and Mathematical Modeling Department. He published two books and over 100 papers. His research interests include adaptive and robust control, signal processing, machine learning and computational intelligence. He is a Member of IEEE, New York Academy of Sciences, Ukrainian National Committee of Automatic Control; he is Ukrainian State Prize in Science and Technology winner.

7

Odor classification by neural networks[*]

Sigeru Omatu

Faculty of Engineering, Osaka Institute of Technology, Japan

Abstract

Metal oxide semiconductor gas (MOG) sensors and quartz crystal microbalance (QCM) sensors are used to measure several kinds of odors. Using neural networks to classify the measured data of odors, artificial electronic noses have been developed. This chapter is to consider an array sensing system of odors and to adopt a layered neural networks for classification. Furthermore, we consider mixed effect of odors for classification accuracy. For simplicity, we will treat the case that two kinds of odors are mixed, since more than two becomes too complex to analyze the classification efficiency. In order to consider the mixed effect, we use as the test data two out of four kinds of odors. An acceptable result, although not perfect, has been achieved for the classification of mixed odors, by using the layered neural network.

Keywords: odor classification, odor sensors, neural networks, layered neural network, features of odor.

7.1 Introduction

It is important to treat smell information in order to achieve the high quality of information like human being since the smell is one of five senses and hu-

---

[*]This work was supported by JSPS KAKENHI Grant-in-Aid for Scientific Research(B)(23360175).

V. Haasz and K. Madani (Eds.), Advanced Data Acquisition and Intelligent Data Processing, 159–179.

man being uses those five senses to communicate and understand each other. Artificial odor sensing and classification systems through electronic technology are called an electronic nose and they have been developed according to various odor sensing systems and several classification methods [1–4].

We have developed electronic nose systems to classify the various odors under different densities using a layered neural network or a competitive neural network of the learning vector quantization method [5–7]. Based on our experience, we have developed a new measurement system such that precise evaluation of the odor can be done with many sensors and by controlling dry air flow rate, temperature, and humidity. Furthermore, we have attached a sensing system with both metal oxide semiconductor gas (MOG) sensors array and quartz crystal microbalance (QCM) sensors array.

After a brief survey of the electronic nose and its measurement and classification methods, we consider the electronic nose accuracy when two odors are mixed after each of the original odors has been classified precisely by using a neural network. We will consider the classification of mixed odors based on the sensing data by using QCM sensors. The QCM sensors are more sensitive than MOG sensors to some kinds of odor and the environmental condition for sensing odors. Using many QCM sensors, we will try to separate mixed odors into original odors by using a neural network.

7.2 Human olfactory processes

Although the human olfactory system is not fully understood by physicians, the main components about the anatomy of human olfactory system are the olfactory epithelium, the olfactory bulb, the olfactory cortex, and the higher brain or cerebral cortex, as shown in Fig. 7.1.

The first process of human olfactory system is to breathe or to sniff the smell into the nose, as shown in Step 1 of Fig. 7.1. The difference between the normal breath and the sniffing is the quantity of odorous molecules that flows into the upper part of the nose. In case of sniffing, most air is flown through the nose to the lung and about 20% of air is flown to the upper part of the nose and detected by the olfactory receptors. In case of sniffing, the most air flow directly to the upper part of the nose interacts with the olfactory receptors. The odorous molecules are dissolved at a mucous layer before interacting with olfactory receptors in the olfactory epithelium, as shown in Step 2 of Fig. 7.1.

The concentration of odorous molecules must be over the recognition threshold. After that, the chemical reaction in each olfactory receptor pro-

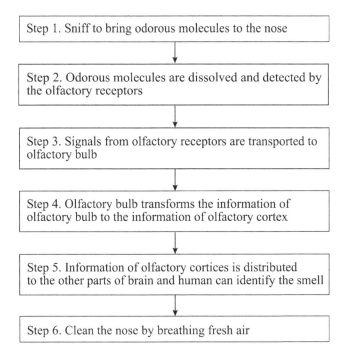

Step 1. Sniff to bring odorous molecules to the nose

Step 2. Odorous molecules are dissolved and detected by the olfactory receptors

Step 3. Signals from olfactory receptors are transported to olfactory bulb

Step 4. Olfactory bulb transforms the information of olfactory bulb to the information of olfactory cortex

Step 5. Information of olfactory cortices is distributed to the other parts of brain and human can identify the smell

Step 6. Clean the nose by breathing fresh air

Figure 7.1 Olfactory system

duces an electrical stimulus. The electrical signals from all olfactory receptors are transported to olfactory bulb, as shown in Step 3 of Fig. 7.1.

The input data from olfactory bulbs are transformed to be the olfactory information to the olfactory cortex, as shown in Step 4 of Fig. 7.1. Then the olfactory cortex distributes the information to other parts to the brain and human can recognize odors precisely, as shown in Step 5 of Fig. 7.1. The other parts of the brain that link to the olfactory cortex will control the reaction of the other organ against the reaction of that smell. When human detects bad smells, human will suddenly expel those smells from the nose and try to avoid breathing them directly without any protection. This is a part of the reaction from the higher brain.

Finally, the cleaning process of the nose is to breathe fresh air in order to dilute the odorous molecules until those concentrations are lower than the detecting threshold, as shown in Step 6 of Fig. 7.1. The time to dilute the smell depends on the persistence qualification of the tested smell.

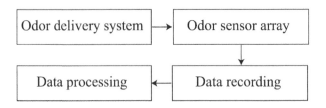

Figure 7.2 Main parts of electronic nose systems

7.3 Electronic nose system

The electronic nose system is an alternative method to analyze smell by imitating the human olfactory system. In this section, the concept of an electronic nose is explained. Then various sensors for odors applied as the olfactory receptors are explained. Finally, the mechanism of a simple electronic nose that will be developed in this chapter is described in detail by comparing the function of each part with the human olfactory process. The mechanism of electronic nose systems can be divided into main four parts as shown in Fig. 7.2.

7.3.1 Odor delivery system

The first process of the human olfactory system is to sniff the odorous molecule into the nose. Thus, the first part of the electronic nose system is the mechanism to bring the odorous molecules into the electronic nose system. There are three main methods to deliver the odor to the electronic nose unit, sample flow, static system, and pre-concentration system.

The sample flow system is the most popular method to deliver odorous molecule to the electronic nose unit. Some carrier gas such as air, oxygen, nitrogen, and so on, is provided as a carrier gas at the inlet port to flow the vapor of the tested smell through the electronic nose unit via the outlet port. The mechanism to control the air flow of an electronic nose may contain various different parts such as a mass flow controller to control the pressure of the carrier gas, a solenoid valve to control the flow of inlet and outlet ports, a pump to suck the tested odor from the sampling bag in case that the tested odor is provided from outside, a mechanism to control humidity, and so on. Most commercial electronic noses contain complicated odor delivering systems and this makes the price of the electronic noses become expensive.

The static system is the easiest way to deliver odorous molecules to the electronic nose unit. The electronic nose unit is put into a closed loop container. Then an odor sample is injected directly to the container by a syringe. It is also possible to design an automatic injection system. However, the rate to inject the test odors must be controlled to obtain accurate results. Normally, this method is applied for the calibration process of the electronic nose. But, in this case, the quantity of the odor may not be enough to make the sensor reach the saturation stage, that is, the stage that sensor adsorbs the smell fully.

The pre-concentration system is used in case of the tested smell that has a low concentration and it is necessary to accumulate the vapor of the tested odor before being delivered to the electronic nose unit. The pre-concentrator must contain some adsorbent material such as silica and the tested odor is continuously accumulated into the pre-concentrator for specific time units. Then, the pre-concentrator is heated to desorb the odorous molecule from the adsorbent material. The carrier gas is flown through the pre-concentrator to bring the desorbed odorous molecules to the electronic nose unit. By using this method, some weak smells can be detected by the sensor array in the electronic nose unit.

7.3.2 Odor sensor array

The second process of the human olfactory system is to measure various odors corresponding to various receptors in the human olfactory system. In order to realize many receptors artificially, we adopted two types of sensors. One is MOG type [6] and the other is QCM type [7]. In this chapter, we use many sensors which are allocated in an array structure for each type of MOG and QCM types as shown in Fig. 7.3. This structure is adopted based on the human olfactory system. As we will explain in what follows, the odor sensors are not so small, which results in a large space to measure odors.

7.3.3 Data recording

The data recording is corresponding to temporal memory for the human olfactory system. In the latter case, after learning odors we could identify an odor suddenly, we store sensing data of odors in a computer. To make reading and writing the data, we make an efficient structure of data base.

7.3.4 Data processing

Using the data base of odors, we must apply an intelligent signal processing technique to recognize odors correctly. We make pre-processing the odor data such as noise reduction, normalization, feature extraction, etc. Then we use

Figure 7.3 Sensor arrray system and a permeater device which is a gas generator for calibration purposes that utilises permeation and diffusion tubes to generate the calibration gas

neural networks for classification of odors. Basically, we use a layered neural networks and competitive networks for odor classification since learning ability and robustness are important in odor classification. The most difficult and important process in the odor classification is to find excellent features which are robust for environment like temperature, humidity, and density levels of odors.

7.4 Principle of odor sensing

Nowadays, there are many kinds of sensors that can measure odorous molecules. However, only a few kinds of them have been successfully applied as artificial olfactory receptors in commercial electronic noses. We show two types of odor sensors which are major for odor sensing. One is a MOG sensor and the other is a QCM sensor, which will be explained in what follows.

7.4.1 Principle of MOG sensors

MOG sensors are the most widely used sensors for making an array of artificial olfactory receptors in electronic nose systems. These sensors are commercially available as the chemical sensor for detecting some specific smells. Generally, an MOG sensor is applied in many kinds of electrical appliances such as a microwave oven to detect the food burning, an alcohol breathe checker to check the drunkenness, an air purifier to check the air quality, and so on.

Figure 7.4 MOG sensors

The picture of some commercial MOG sensors is shown in Fig. 7.4. Various kinds of metal oxides, such as SnO_2, ZnO_2, WO_3, TiO_2 are coated on the surface of a semiconductor. But, the most widely applied metal oxide is SnO_2. These metal oxides have a chemical reaction with the oxygen in the air and the chemical reaction changes when the adsorbing gas is detected. The scheme of chemical reaction of an MOG sensor when adsorbing with the CO gas, is shown as follows:

$$\frac{1}{2}O_2 + (SnO_{2-x}) \rightarrow O^-\text{ad}\ (SnO_{2-x}) \tag{7.1}$$

$$CO + O^-\text{ad}\ (SnO_{2-x}) \rightarrow CO_2 + (SnO_{2-x}) \tag{7.2}$$

When the metal oxide element on the surface of the sensor is heated at a certain high temperature, the oxygen is adsorbed on the crystal surface with the negative charge as shown in Fig. 7.5. In this stage, the grain boundary area of the metal oxide element forms a high barrier as shown in the left hand side of Fig. 7.5.

Then, the electrons cannot flow over the boundary and this makes the resistance of the sensor become higher. When the deoxidizing gas, e.g., CO gas, is presented to the sensor, there is a chemical reaction between negative charges of oxygen at the surface of the metal oxide element and the deoxidizing gas as shown in (7.1). The chemical reaction between adsorbing gas and the negative charge of the oxygen on the surface of MOG sensor reduces the grain boundary barrier of the metal oxide element as shown in the right hand side of Fig. 7.5. Thus, the electron can flow from one cell to another cell easier. This makes the resistance of MOG sensor lower by the change of oxygen pressure according to the rule of (7.2). Thus, (7.1) means that CO_2 is reduced and (7.2) means that CO is oxidized.

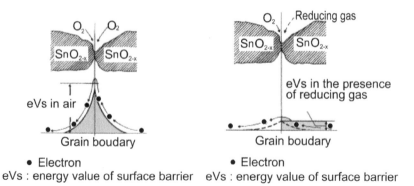

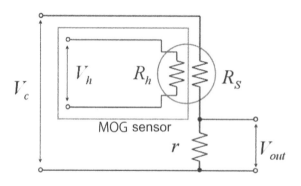

Figure 7.5 Principle of MOG sensor

Figure 7.6 Output voltage of MOG sensors

The relationship between sensor resistance and the concentration of deoxidizing gas can be expressed by the following equation over certain range of gas concentration:

$$R_s = A[C]^{-\alpha} \qquad (7.3)$$

where R_s =electrical resistance of the sensor, A = constant, C = gas concentration, and α =slope of R_s curve. The electric circuit for the MOG sensor is shown in Fig. 7.5. Electrical voltages are provided to the circuit (V_c) and the heater of the sensor (V_h). When the MOG sensor is adsorbed with oxygen and the deoxidizing gas, the resistance of the sensor (R_s) is changed. Thus, we can measure the voltage changes (V_{out}) while the sensor is adsorbing the tested odor.

Table 7.1 List of MOG sensors from the FIS Inc. used in this experiment

Sensor Model	Main Detecting Gas
SP-53	Ammonia, Ethanol
SP-MW0	Alcohol, Hydrogen
SP-32	Alcohol
SP-42A	Freon
SP-31	Hydrocarbon
SP-19	Hydrogen
SP-11	Methane, Hydrocarbon
SP-MW1	Cooking vapor

MOG sensors need to be operated at high temperature, so they consume a little higher power supply than the other kinds of sensors. The reliability and the sensitivity of MOG sensors are proved to be good to detect volatile organic compounds (VOCs), combustible gas, and so on. However, the choices of MOG sensors are still not cover all odorous compounds and it is difficult to create an MOG sensor that responds to one odor precisely. Generally, most commercial MOG sensors respond to various odors in different ways. Therefore, we can expect if we use many MOG sensors to measure a smell, the vector data reflect the specific properties for the smell.

Generally, it is designed to detect some specific smell in electrical appliances such as an air purifier, a breath alcohol checker, and so on. Each type of MOG sensors has its own characteristics in the response to different gases. When combining many MOG sensors together, the ability to detect a smell is increased. The main part of the MOG sensor is the metal oxide element on the surface of the sensor. When this element is heated at certain high temperature, the oxygen is adsorbed on the crystal surface with the negative charges. The reaction between the negative charges of the metal oxide surface and deoxidizing gas makes the resistance of the sensor vary as the partial pressure of oxygen changes. Based on this characteristic, we can measure the net voltage changes while the sensors adsorb the tested odor.

One of sample data of MOG sensors has been shown in Fig. 7.8 where two trials have been performed. In the first trial, at 2,200 [s] a gas is inlet and in the second trial, at 3,800 [s] a little bit thinner gas is inlet. There are many kinds of features to discriminate each gas from the others. From practical view point, five dimension(three slopes in each segment plus the maximum value and the mean value among the total data) as shown in Fig. 7.7. For simplicity, we adopt only the maximum value. From Fig. 7.8 we can see that maximum values have been decreased at the second trial compared with the

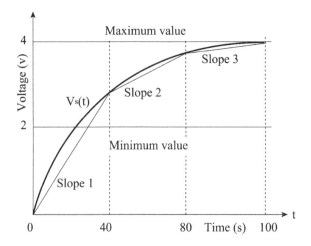

Figure 7.7 Features of MOG sensors

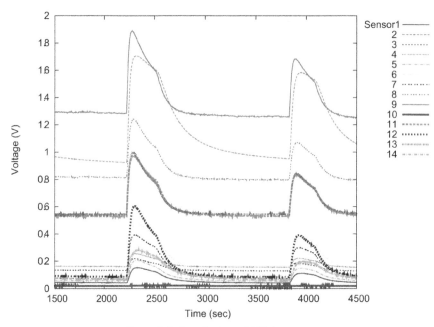

Figure 7.8 Sample data of MOG sensors

first trial which means that the MOG sensors depend on the density levels of the gas.

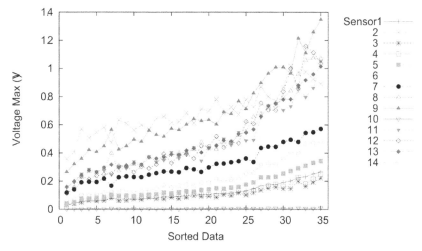

Figure 7.9 Sorting results of the maximum values for sensors

We have rearranged the maximum values for each trial according to as-cending order and the results become Fig. 7.9. Since the resulting figure might be modelled by logarithmic fitting, we have transformed the density levels and fitted linear regression lines. The results are shown in Fig. 7.10 and the slopes are independ on the densities. Therefore, the slopes are one of the important factor which discriminate the kinds of odors without depending their densities.

7.4.2 Principle of QCM sensors

QCM sensors have been well-known to provide a very sensitive mass-measuring devices in Nano-gram levels, since the resonance frequency will change upon the deposition of the given mass on the electrodes. Synthetic polymer-coated QCM sensors have been studied as sensors for various gasses since a QCM sensor is coated with a sensing membrane works as a chemical sensor. The QCM sensors are made by covering the surface with several kinds of a very thin membrane with about 1 μm as shown in Fig. 7.10.

Since the QCM sensor oscillates with a specific frequency depending on the cross section corresponding to three axis of the crystal, the frequency will change according to the deviation of the weight due to the adsorbed odor molecular (odorant). The membrane coated on QCM sensor has selective ad-sorption rate for a molecular and the frequency deviation show the existence

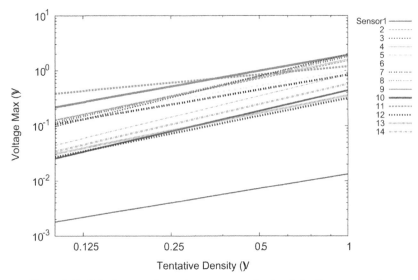

Figure 7.10 Odor features for various concentration levels by linear regression

of odorants and their densities. Odorants and membranes are tight relation while it is not so clear whose materials could be adsorbed so much.

In this chapter, we have used the following materials as shown in Table 7.2. The reason why fluorine compounds are used here is that the compounds repel water such that pure odorant molecular could be adsorbed on the surface of the membrane. To increase the amount of odorants to be adsorbed it is important to iron the thickness of the membrane. In Table 7.2, we have tried to control the density of the solute in the organic solvent. The basic approach used here is a sol-gel method. The sol-gel process is a wet-chemical technique used for the fabrication of both glassy and ceramic materials. In this process, the sol (or solution) evolves gradually towards the formation of a gel-like network containing both a liquid phase and a solid phase. Typical precursors are metal oxides and metal chlorides, which undergo hydrolysis and poly-condensation reactions to form a colloid. The basic structure or morphology of the solid phase can range anywhere from discrete colloidal particles to continuous chain-like polymer networks.

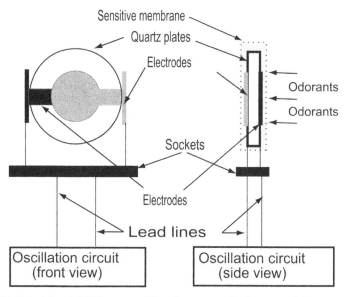

Figure 7.11 Principle of QCM sensors. The odorants attached on a sensitive membrane will make the weight change of quartz plane. Thus, the original frequency of the crystal oscillation will become smaller according to the density of odorants

Table 7.2 Chemical materials used as the membrane where E:etanol, W:water, DA:dilute nitric acid, EA:ethyl acrylate, MTMS:trimethoxy silane, PFOEA:perfluorooctylethyl acrylate

Sensor number	Materials of membrane
Sensor 1	E(4ml), DA(0.023ml),
Sensor 2	W(3.13ml), E(4ml), EA(0.043ml), MTMS
Sensor 3	W(3.13ml), E(4ml), EA(0.014ml), MTMS, PFOE
Sensor 4	W(3.13ml), E(4ml), EA(0.015ml), MTMS, PFOE
Sensor 5	W(0.30ml), E(4ml), EA(0.043ml), MTMS, PFOE
Sensor 6	W(0.05ml), E(3.0ml), EA(0.043ml), MTMS, PFOE
Sensor 7	W(0.30ml), E(3.2ml), EA(0.043ml), MTMS, PFOE
Sensor 8	No membrane

7.5 Odor sensing system

Generally, odor sensors are designed to detect some specific odor in electrical appliances such as an air purifier, a breath alcohol checker, and so on. Each of odor sensors such as MOG sensors or QCM sensors has itself characteristics in the response to different odors. When combining many odor sensors together, the ability to detect an odor is increased. An electronic nose system

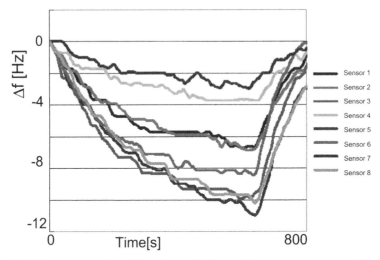

Figure 7.12 Measurement data of Experiment III. Here, eight sensors are used and for 800 [s] the data were measured. The maximum value for each sensor among eight sensors is selected as a feature value for the sensor. Therefore, we have eight dimensional data for an odor and they will be used for classification

shown in Fig. 7.13 has been developed, based on the concept of human olfactory system. The combination of odor sensors, listed in Tables 7.1 and 7.2, are used as the olfactory receptors in the electronic nose.

7.6 Classification method of odor data

In order to classify the odors we adopt a three-layered neural network based on the error back-propagation method as shown in Fig. 7.15. The error back-propagation algorithm which is based on a gradient method is given by the following steps.

Step 1. Set the initial values of $w_{ji}, w_{kj}, \theta_j, \theta_k$, and $\eta(> 0)$.

Step 2. Specify the desired values of the output $d_k, k = 1, 2, \ldots, K$ corresponding to the input data $x_i, i = 1, 2, \ldots, I$ in the input layer.

Step 3. Calculate the outputs of the neurons in the hidden layer and output layer by

$$\text{net}_j = \sum_{i=1}^{I} w_{ji} x_i - \theta_j, O_j = f(\text{net}_j), f(x) = \frac{1}{1 + e^{-x}} \qquad (7.4)$$

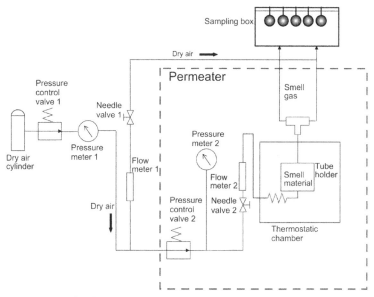

Figure 7.13 Odor sensing framwork. The air will be emitted from the dry air cylinder. Air flow is controlled by pressure control valves 1 and 2. By using the needle valve 2, more precise flow rate of the dry air can be achieved and the thermostatic chamber in the permeater can control the temperature of the dry air. Finally, the air is pull in the sampling box where the MOG sensors and/or QCM sensors are attached on the ceiling of the box

$$\text{net}_k = \sum_{j=1}^{J} w_{kj} O_j - \theta_k, O_k = f(\text{net}_k). \tag{7.5}$$

Step 4. Calculate the error e_k and generalized errors δ_k ,δ_j by

$$e_k = d_k - O_k, \delta_k = e_k O_k (1 - O_k) \tag{7.6}$$

$$\delta_j = \sum_{k=1}^{K} \delta_k w_{kj} O_j (1 - O_j). \tag{7.7}$$

Step 5. If e_k is sufficiently small for all k, END and otherwise

$$\Delta w_{kj} = \eta O_j \delta_k, \quad w_{kj} \Leftarrow w_{kj} + \Delta w_{kj} \tag{7.8}$$

$$\Delta w_{ji} = \eta O_i \delta_j, \quad w_{ji} \Leftarrow w_{ji} + \Delta w_{ji}. \tag{7.9}$$

Step 6. Go to *Step* 3. Using the above recursive procedure, we can train the odor data. The measurement data is an eight-dimensional vector which are obtained with eight sensors stated in Table 7.2.

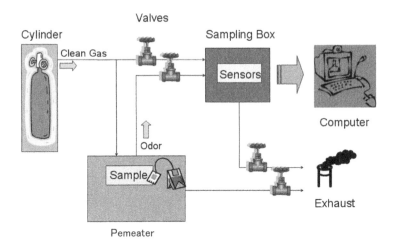

Figure 7.14 Odor sensing systems

Table 7.3 Teas used in Experiment I

Label	Materials	Samples
A	English tea	20
B	Green tea	20
C	Barley tea	20
D	Oolong tea	20

7.7 Classification results using MOG sensors

We have measured four types of tea using MOG sensors shown in Table 7.1. The odors used here are shown in Table 7.3. Note that the chemical properties of these odors are very similar and it has been difficult to separate them based on the measurement data by using MOG sensors. We have examined two examples, Experiments I and II. For Experiment I, we classify kinds of teas. The numbers of neurons for this experiment are eight in the input layer, four in the hidden layer, and four in the output layer, that is, 8-4-4 structure.

The number of training samples is fifteen and the number of test samples is five. We change the training data set for 100 times and check the classification accuracy for the test data samples. Thus, we have obtained 500 test samples as the total number of classification. The classification results are summarized in Table 7.4. Average of the classification is 96.2%.

For Experiment II, we consider to classify the five kinds of coffees as shown in Table 7.5 where the smell data A, B, and C are the coffees of Mocha

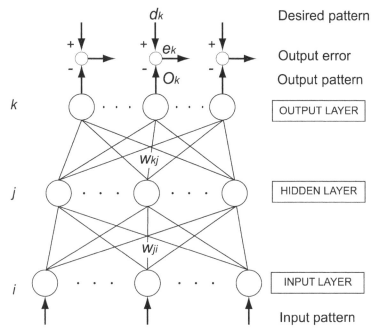

Figure 7.15 Three layered neural network with the error back-propagation. The neural network consists of three layers, that is, an input layer i, a hidden layer j, and an output layer k. When the input data $x_i, i = 1, 2, \ldots, I$ are applied in the input layer, we can obtain the output O_k in the output layer which is compared with the desired value d_k which is assigned in advance. If the error $e_k = d_k - O_k$ occurs, Then, the weighting coefficients w_{ji}, w_{kj} are corrected such that the error becomes smaller based on an error back-propagation algorithm

Table 7.4 Classification results for Experiment I

Odor data	Classification results(96.2%)					
	A	B	C	D	Total	Correct
A	500	0	0	0	500	100.0%
B	0	494	6	0	500	98.8%
C	0	71	429	0	500	85.8%
D	0	0	0	500	500	100.0%

made from different companies. The numbers of neurons are eight in the input layer, five in the hidden layer, and five in the output layer, that is, 8-5-5 structure.

The numbers of training samples are twenty and he number of test samples is fifteen. We change the training data set for hundred times and

Table 7.5 Smell data of teas used in Experiment II

Label	Materials	No. of samples
A	Mocha coffee1	35
B	Mocha coffee2	35
C	Mocha coffee3	35
D	Kilimanjaro coffee	35
E	Char-grilled coffee	35

Table 7.6 Classification results for Experiment II

Odor data	Classification results(88.8%)						
	A	B	C	D	E	Total	Correct
A	1190	253	29	27	1	1500	79.3%
B	225	1237	9	10	19	1500	82.5%
C	142	7	1325	26	0	1500	88.3%
D	9	14	3	1437	37	1500	95.8%
E	0	18	0	11	1471	1500	98.1%

check the classification accuracy for the test data samples. Thus, we have obtained 1,500 test samples as the total number of classification. The classification results are shown in Table 7.6. From Table 7.6 we can see that the total classification is 88.8%. Compared with Table 7.4, this case is worse by about 7%. In the latter case, the classification of Mocha coffees 1, 2, and 3 is not so good. The difference of the company may affect the classification results in a bad way although the smells are similar.

7.8 Classification results using QCM sensors for mixed odor data

We have measured four types of odors, as shown in Table 7.7. The sampling frequency is 1 [Hz], the temperatures of odor gases are 24~26 [°C], and the humidity of gas is 6~8%. To control the density of gases, we use diffusion tubes. Odor data are measured for 600 [s]. They may include impulsive noises due to the typical phenomena of QCM sensors. We call this experiment as Experiment III.

To remove these impulsive noises we adopt a median filter which replaces a value at a specific time by a median value among neighboring data around the specific time. In Fig. 7.12, we show the measurement data for the symbol A (ethanol) where the horizontal axis is the measurement time and the ver-

Table 7.7 Kinds of odors measured of Experiment III

Symbols	Kind of odors
A	Ethanol
B	Water
C	Methyl-salicylate
D	Triethyl-amine

Table 7.8 Training data set for ethanol (A), water (B), methyl-salicylate (C), and tri-ethylamine (D) for Experiment III

Symbols	Output A	Output B	Output C	Output D
A	1	0	0	0
B	0	1	0	0
C	0	0	1	0
D	0	0	0	1

tical axis is the frequency deviation from the standard value (9 [MHz]) after passing through a five-point median filter.

7.8.1 Training for classification of odors

In order to classify the feature vector, we allocate the desired output for the input feature vector where it is nine-dimensional vector, as shown in Table 7.8 since we have added the coefficient of variation to the usual feature vector to reduce the variations for odors. The training has been performed until the total error becomes less than or equal to 0.5×10^{-2} where $\eta = 0.3$.

7.8.2 Classification results and discussion for mixed odor data

After training such that all the smells, A, B, C, and D, have been classified correctly, we have tested the mixed data sets such that two kinds of odors are mixed with the same rate where the data set of mixed smells are {A&B, B&C, C&D, D&A, A&C, B&D}. Then, the classification results are shown in Table 7.9 where the values in boldface denote the top case where the maximum output values are achieved. The maximum values show one of the mixed odors. But, some of them do not show the correct classification for the remaining odor. Thus, we have modified the input features such that

$$z = x - 0.9y \tag{7.10}$$

Table 7.9 Testing the mixed odors for Experimet III. Here, bold face denotes the largest value

Symbols	Output A	Output B	Output C	Output D
A and B	**.673**	.322	.002	.001
B and C	.083	**.696**	.174	.001
C and D	.001	.004	.016	**.992**
D and A	.003	.003	.002	**.995**
A and C	**.992**	.006	.002	.000
B and D	.003	.003	.003	**.995**

Table 7.10 Testing the mixed odors where except for the largest value the top is selected as the second odor among the mixed odors for Experiment III. Here, asterisk * denotes the top except for the largest value

Symbols	Output A	Output B	Output C	Output D
A and B	.263	**.290***	.166	.066
B and C	.358	.029	**.631***	.008
C and D	.002	.071	**.644***	.163
D and A	**.214***	.004	.037	.230
A and C	.031	.020	**.527***	.026
B and D	.108	.010	**.039***	.325

where x is the feature, y denotes the top value of each row in Table 7.9, and z is a new feature. Using the new feature vector, we have obtained the classification results as shown in Table 7.10. By changing the features according to the above relation, better classification results have been obtained. But, the coefficient 0.9 used in the above equation is not necessarily appropriate. The value might be replaced by the partial correlation coefficient in multivariate analysis.

7.9 Conclusions

In this chapter, a new approach to odor classification has been presented and discussed by using MOG sensors and QCM sensors. After surveying the smell sensing and classification methods, we have examined two examples, Experiment 1 and Experiment II. Then, mixed effects of two different odors have been considered in Experiment III. From these results, we could know that the electronic nose systems might be applied to many applications in real world such that small detection of bad gas or uncomfortable smell, Odor classification of perfume of joss tics kinds, food freshness, etc.

References

[1] Milke J. A. Application of neural networks for discriminating fire detectors. In *International Conference on Automatic Fire Detection*, 1995.

[2] Siciliano P. Distante C., Ancona N. and Burl M. Support vector machines for olfactory signal recognition. *Sensors and Actuators B*, 88(1):30–39, 2003.

[3] BGardner J. and Bartlet P. *Electronic Noses: Principles and Applications.* Oxford University Press, 1998.

[4] Bicego M. Odor classification using similarity-based representation. *Sensors and Actuators B*, 110(2):225–230, 2005.

[5] S. Omatu. *Pattern Analysis for Odor Sensing System.* IGI Global, 2012.

[6] S. Omatu and M. Yano. Intelligent electronic nose system independent on odor concentration. In *Proceedings of the Second International Workshop on User Modeling*, 2011.

[7] M. Yoshioka T. Fujinak and T. Kosaka. Intelligent electronic nose systems for fire detection systems based on neural networks. In *Proceedings of the Second International Conference on Advanced Engineering Computing and Applications in Sciences*, 2008.

Biography

Sigeru Omatu was born on December 16, 1946. He received his Ph.D. in Electronic Engineering from Osaka Prefecture University and joined the faculty at University of Tokushima in 1969. He was Professor of University of Tokushima in 1988 and Professor of Osaka Prefecture University in 1995. He has been Professor of Osaka Institute of Technology since 2010. His honors and awards include the Best Paper Awards of IEE of Japan, 1991, SICE, Japan, 1995. His research area is intelligent signal processing based on neural networks.

8

ANFIS based approach for improved multisensors signal processing

Nataliya Roshchupkina[1], Anatoliy Sachenko[2,3], Oleksiy Roshchupkin[4], Volodymyr Kochan[2] and Radislav Smid[4]

[1]*Chernivtsi National University, Chernivtsi, Ukraine,*
[2]*Ternopil National Economic University, Ternopil, Ukraine*
[3]*Silesian University of Technology, Gliwice, Poland*
[4]*Czech Technical University in Prague, Czech Republic*

Abstract

An improved algorithm of intelligent data processing by integrating the modified identification method of individual characteristic curve along with the adaptive neuro-fuzzy inference system (ANFIS) was proposed. The results of data prediction with training the ANFIS system and different number of training epochs are described. The proposed approach provides the high accuracy of the data prediction as well as a low level of neural network training error.

Keywords: adaptive neuro-fuzzy inference system, multisensor, neural network, identification, prediction, individual characteristic curve, correction, calibration

8.1. Introduction

Multisensors (MS) [1] have gained wide applications in recent years. They are able to measure simultaneously several physical quantities. An output signal for traditional sensors depends mainly on a single measurable physical quantity. A dependence of the output signal not only from the measured physical quantity is considered a disadvantage that causes the sensor error [26]. The MS output signal purposefully depends on several physical quantities. Usually MS are produced in thin-film and semiconductor integrated technology and they are used in the chemical industry, various security systems, environmental monitoring in particular

V. Haasz and K. Madani (Eds.), Advanced Data Acquisition and Intelligent Data Processing, 181–211.

ultraviolet sensors, and in other fields. In these fields, they have significant advantages such as an ability to measure many physical quantities, which often couldn't be measured by other sensors, simplicity and relatively low price.

MS also have some imperfections. First, processing of the MS output signal must separate out the values of individual measured quantities. If these quantities have influence on the same parameter of MS output signal [2, 3], so it's often necessary to use methods of identification based on artificial intelligence methods [4]. The neural network (NN) methods for recognition have shown best results in these fields of applications. This is due to good adaptive and generalizing properties of NN [5-7].

Second, available on the market MS have significant deviations of individual characteristic curve (ICC) in comparison with the nominal [2]. Often the nominal characteristic curve of MS is specified with low precision (sometimes by a graph) [2]. Therefore, usually measuring devices that use MS have low accuracy.

Methods to improve accuracy of measurement instruments and systems with MS, usually based on the turning to ICC that is defined by the results of experimental studies of MS (their calibration) [8]. It is usually considered two ways of using the results of calibration: to make a selection MS with specified characteristic curve and other way – to make a correction of ICC deviations from the nominal. The first way is used rarely. It's characterized by a considerable complexity because of necessity experimentally to conduct research of all MS and usually the percentage of selected MS is a small one even for ordinary, one-parameter sensors. For MS it's necessary to provide the specified allowable deviation from the nominal characteristic curve for all physical quantities. Therefore, the percentage of selected MS is determined as a product of the probabilities of selection by all physical quantities, so it will be quite small.

Much more prospective approach is the correction of deviations ICC from the nominal. This method also requires an experimental research of all MS [12], but almost all MS can be used. However, for MS it is necessary to conduct a research study their ICC for measured physical quantities involved. Let's estimate the required number of points for MS calibrating by analogy with conventional sensors. For example, for resistance thermometers (this is a one-parameter sensor) the nominal characteristic curve is defined with high precision [9, 26]. The sensors themselves have small deviations from the nominal characteristic curve. To identify its ICC it is sufficiently $k \approx 2...3$ calibration points (see Fig. 8.1b) (depending on the desired accuracy and desired width measurement range with high accuracy). The S_1, S_2, ... S_n are input values, and n_1, n_2, ... n_n, are corresponding output values.

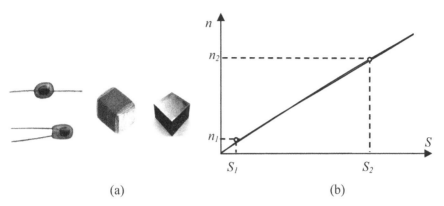

(a)   (b)

Fig. 8.1 (a) Glass coated axial and radial bead thermistors; (b) Recommended number of calibration points for the resistance thermometers is about $k \approx 2...3$

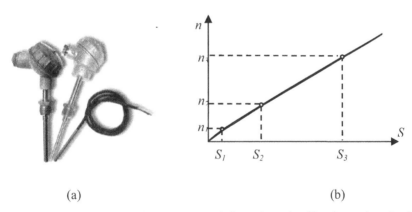

(a)   (b)

Fig. 8.2 (a) Thermocouples; (b) Recommended number of calibration points for the standard thermocouples is about $k \approx 3...4$

For the standard thermocouples the nominal characteristic curve has lower accuracy and the recommended number of calibration points is about $k \approx 3...4$ (see Fig. 8.2b), and for non-standard thermocouples the calibration points is about $k \approx 5...7$ (see Fig. 8.3) [9, 26].

The latter value can be considered as acceptable for MS. It should be noted that for MS the changing of one physical quantity usually affects the MS sensitivity to other physical quantities, i.e. characteristic curve for various physical quantities are interdependent. Therefore, for MS the number of calibration points for two measured physical quantities should correspond to the square of the number calibration points one-parameter sensors, i.e. $k=49$.

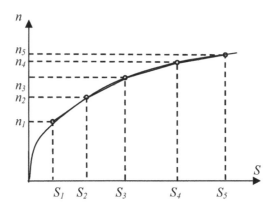

Fig. 8.3 Recommended number of calibration points for the not standard thermocouples is about $k \approx 5...7$

For three measured physical quantities the number of calibration points should meet the cube and higher degree – for the greater number of physical quantities. This dramatically increases the complexity of identifying ICC of MS and the cost of laboratory equipment necessary for its implementation.

8.2. Identification methods

In general the mathematical model of sensor or process expresses the relationships between input signal and output signals. Very often some parameters of the process model may be unknown as well as a model itself. During the process, the faults of the models have to be rather precise in order to express deviations as result of process faults. But also the identification method itself may be as a source to gain information, e.g. process parameters that change under the influence of the faults. First publications known on fault detection with identification methods are [27-31].

To reduce the complexity of identifying ICC of MS the approximation (interpolation and extrapolation) can be used for calibration results in a reduced number of points. In mathematics there are many known methods of function approximation, which provide high accuracy with a small number of points – a polynomial approximation [11], approximation by rational functions (using Floater-Hormann basis) [12], approximation splines (spline regression with a penalty function) [13], approximation using linear least squares method [14]. But for the approximation of the surface (e.g., two-parameter sensor) universal mathematical method is to split the surface on a set of elementary flat rectangles or squares with so small sizes that the non-linearity of sensor characteristic curve can be neglected within limits of these squares. It is clear that significant improving the accuracy of MS is required the number of calibration points even greater than $k=49$ for this approximation. Many processes show a nonlinear static and dynamic behaviour [22], especially in wide areas of operations for example vehicles, aircraft, combustion engines and thermal plants and all processes with Coulomb friction and magnetic hysteresis. Classical nonlinear models in combination with parameter estimation methods as well as model architectures originating from the field of artificial NN are well suited to the identification of nonlinear static and dynamic processes.

To reduce the number of calibration points for two-parameter sensor by approximation of its characteristic curve is proposed to use NN [15], and even the support vector method (Support Vector Machine) [16]. It allowed to decrease the number of calibration points up to $k=9$. But the obtained errors of the MS characteristic identification at points – where the calibration isn't carried out – is quite large, 0.5%. In addition, the methods [15, 16] were investigated only for conditions that calibration errors are absent. In practice the effect of measurement errors during the calibration of two different physical quantities (for two-parameter sensor) is equivalent to random noise, which significantly increases the approximation error by methods [15, 16].

The completed analysis of identification methods for MS output signals is carried out in [21].Their conclusion is that NN methods give the best performance, but they require some time for training and expert intuition for the meaningful use.

To improve an accuracy it's necessary to use calibration i.e. determine (identify) a difference between a real conversion function and ideal one. In Fig. 8.4 [22] those methods are extracted which can be applied for a wide range of processes and excitation signals at the input.

Especially for dynamic processes when the input signals have to change periodically, stochastically or by special test signals. These signals may be the normal operating signals (like in servo systems, actuators or driving vehicles) or may be artificially introduced for testing (after fault detection with other methods or for quality control during manufacturing or maintenance) [24, 25]. A considerable advantage of identification methods is that with only one input and one output signal several parameters (up to about six) can be estimated, which give a detailed picture on internal process quantities.

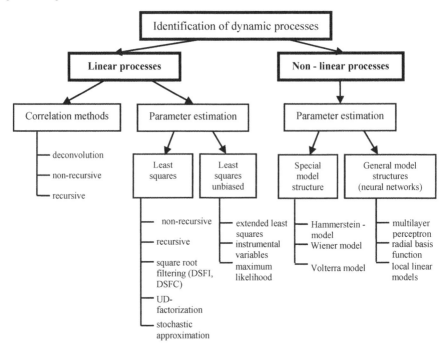

Fig. 8.4 Survey of identification methods [22].

8.2.1 Artificial neural networks for identification

This analysis of identification methods have shown the necessity to develop the method of MS characteristic curve identification and minimize the number of calibration points from k=49 to k=9 (in particular of two-parameter sensors) with low cost experimental studies.

For a general identification approach, methods of interest are those that do not require specific knowledge of the process structure and hence are widely applicable. Artificial NN fulfil these requirements [10, 22]. They are composed of mathematically formulated neurons. At first, these neurons were used to describe the behaviour of biological neurons [32]. The interconnection of neurons in networks allowed the description of relationships between input and output signals, [33, 34]. In the sequel, artificial NN are considered as a map of input signals u to output signals y (Fig. 8.5). Usually, the adaptable parameters of NN are unknown. As a result, they have to be adapted or "trained" or "learned" by processing measured signals u and y [35, 36]. This is a typical system identification problem. If inputs and outputs are gathered into groups or clusters, a classification task in connection with, e.g. pattern recognition is given [37].

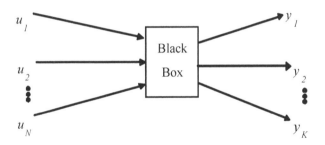

Fig. 8.5 System with N inputs and K outputs, which has to be approximated by an artificial NN

8.2.2 Neuron model

The main component of the NN are neurons (elements, nodes), which are connected together. The signals are distributed using their weighted connections with weighting coefficient or weight. There are software and hardware NN models. The software models are very spread. It can be used for pattern recognition, prediction, approximation, etc.

The input of the artificial neuron receives a set of signals that are outputs of other neurons. Each input is multiplied by the corresponding weight, similar to its synaptic strength. Then all outputs are added together and define the activation level of the neuron. Figure 8.6. illustrates a model that implements this approach. It was a plenty of networking paradigm, but almost all of them are based on this configuration. The set of input signals marked x_1, x_2, ..., x_n, enters the artificial neuron. These input signals, collectively denoted by the vector X. They correspond to the signals that come in to synapses of biological neuron.

Each signal is multiplied by the corresponding weight w_1, w_2, ..., w_n, and enters to adding block labeled Σ. Each weight corresponds to the "strength" of a biological synaptic connections (set of this weights is denoted by the vector W). Adding unit that corresponds to the body of biological neuron add inputs algebraically, produces output signal call NET. In vector representation this can be written compactly as $NET = X \cdot W$.

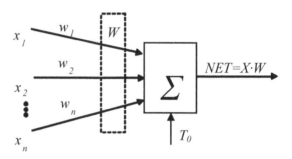

Fig. 8.6 Artificial neuron model

8.2.3 Activation functions

The signal NET then usually is converted by activation function F and gives the output neurons signal $OUT = F\ (NET)$. Activation function F (NET) can be:

1. Threshold binary function (Fig. 8.7a)

$$OUT = \begin{cases} 1, & NET \geq T \\ 0, & NET < T \end{cases},$$
(8.1)

where T is a constant threshold value, or a function that more accurately models the nonlinear transfer characteristic of biological neuron;

2. Linear constraints function (Fig. 8.7b)

$$OUT = \begin{cases} 1, NET \geq T \\ NET, 0 \leq NET < T \\ 0, NET < 0 \end{cases} ; \qquad (8.2)$$

3. Hyperbolic tangent function (Fig. 8.7 c)

$$OUT = th(C \cdot NET) = \frac{\exp(C \cdot NET) - \exp(-C \cdot NET)}{\exp(C \cdot NET) + \exp(-C \cdot NET)}, \qquad (8.3)$$

where $C > 0$ is sigmoid width ratio on the horizontal axis (usually $C = 1$).

4. Sigmoid (*S*-shaped) or logistic function (Fig. 8.7d)

$$OUT = \frac{1}{1 + e^{-C \cdot NET}}. \qquad (8.4)$$

From the expression of sigmoid function obvious that the output value of the neuron is located in the range [0, 1] (Fig. 8.7). The popularity of sigmoid function is determined by its properties as follows: (i) the ability to amplify weak signals better than strong signals; (ii) monotone and differentiation throughout the x-axis; (iii) simple expression for the derivative:

$$F'(NET) = C \cdot F(NET) \cdot (1 - F(NET)). \qquad (8.5)$$

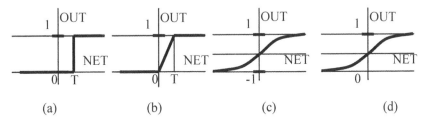

(a)   (b)   (c)   (d)

Fig. 8.7 Activation functions: a) threshold binary; b) linear constraints; c) hyperbolic tangent; d) sigmoid or logistic

The single neurons are connected to a network structure, Figure 8.8. The signal has to go through different layers with parallel arranged neurons: the input layer, the first, second,..., hidden layer and the output layer. With respect to the range internal links values, the input signals can be either binary, discrete or continuous. Binary and discrete signals are used mostly for classification, the continuous signals are used for approximation, prediction, identification tasks, etc. [38, 39].

As approximation NN [40] it is advisable to use a three-level perceptron model (Fig. 8.8).

The perceptron is one of the earliest NN. Invented at the Cornell Aeronautical Laboratory in 1957 by Frank Rosenblatt [33], the perceptron was an attempt to understand human memory, learning, and cognitive processes. In 1960, Rosenblatt demonstrated the Mark I perceptron. The Mark I was the first machine that could "learn" to identify optical patterns.

Neuron of input level (it received value of x) performs distributing functions (see Fig. 8.8). The N neurons of intermediate or hidden level (output values $h_1...h_N$) performs information processing. The neuron of output levels (with output values y) processes information from the intermediate level and returns the result.

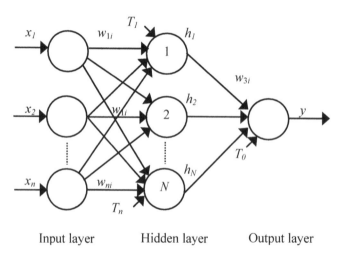

Fig. 8.8 Perceptron structure

The each neuron's output of the pervious level is connected by synapses with inputs of all neurons of the next level. Because of this used perceptron has homogeneous and regular structure. The initial value of output neuron is calculated according to:

$$y = \sum_{i=1}^{N} w_{i3}h_i - T_o,$$ (8.6)

where w_{i3} is the weight coefficient (synapse) from an i neuron of hidden
level to output neuron,

h_i is output values of an i neuron of hidden level,

T_0 isthreshold (offset, bias) of the output neuron [40].

The output values of an i neuron of hidden level is calculated according to:

$$h_i = F(w_{1i}x - T_i), \tag{8.7}$$

where $F(x)$ is neuron's activation function,

w_{1i} is synapse from the output neuron to the i neuron of hidden
level,

T_i is threshold (offset, bias) of an i neuron of hidden level.

For neurons of hidden level is used nonlinear sigmoid activation
function (8.4). For output neuron is used the linear activation function (8.2).

Thus, the approximation NN is a heterogeneous NN in which the
neurons at different levels have different activation functions. To study
approximation NN as heterogeneous NN, it is advisable to use multiple
error propagation algorithm. This algorithm based on modifying of the
algorithm of error backpropagation and provides effective training
heterogeneous NN [39]. Back propagation algorithm based on the method
of gradient descent in the space of synapses and neurons thresholds [41,
42]. For each input training vector p is performed the iterative procedure of
modification for this space, according to:

$$\Delta w_{ij}(t) = -\alpha \frac{\partial E^P(t)}{\partial w_{ij}(t)} \tag{8.8}$$

$$\Delta T_j(t) = -\alpha \frac{\partial E^P(t)}{\partial T_j(t)}, \tag{8.9}$$

where α is a training step,

t for the learning vector p, $p \in \{1, P\}$, P is the size of training
set and

$\dfrac{\partial E^P(t)}{\partial w_{ij}(t)}$, $\dfrac{\partial E^P(t)}{\partial T_j(t)}$ are gradients of an error function for training
iterations.

For each input training vector p the modification of synapses and neuron's
thresholds is processed firstly for the output level neuron (at the time t) and
then to the hidden level neurons (at the time $t+1$).

The training mean square error is calculated according to:

$$E^P(t) = 0.5(y^P(t) - d^P)^2, \qquad (8.10)$$

where $y^P(t)$ is an output value of NN,

d^P is the required value on iteration t for each training vector p. During training process the general error of NN is decrease according to

$$E(t) = \sum_{p=1}^{P} E^P(t), \qquad (8.11)$$

where $p \in \{1, P\}$, P is a size of training set.

In order to improve the training parameters of NN and deficiencies of the classical algorithm of back propagation associated with the empirical choice of constant training step, it is advisable to use the method of steepest descent [39, 40] to calculate the adaptive training step according to:

$$\alpha_3^P(t) = 1/(1 + \sum_{i=1}^{N} h_i^P(t)), \qquad (8.12)$$

for the output neuron with linear activation function (8.2), and:

$$\alpha_2^P(t) = 4/(1 + (x^P)^2) \sum_{i=1}^{N} (w_{i3}^P(t))^2 h_i^P(t)(1 - h_i^P(t)), \qquad (8.13)$$

for neurons in the hidden level with sigmoid activation function (8.4).

8.2.4 Adaptive neuro-fuzzy inference system for identification

For MS ICC identification, the method [43, 18] can be used. However, some practical applications require the faster training and adaptation to the exploitation conditions [19, 20], so it is suitable to combining the modified method of ICC identification [43, 18], with the Adaptive Neuro-Fuzzy Inference System (ANFIS) [44].

The performance of ANFIS method is like both NN and fuzzy logic (FL). In both NN and FL methods, the input passes through the input layer (using input membership function) and the output could be seen in output layer (output membership functions). Since, in this type of advanced fuzzy logic, NN has been used, therefore, by using a learning algorithm the parameters have been changed until reach the optimal solution. The FL tries by using advantages of the NN to adjust its parameters. The ANFIS uses either backpropagation or a combination of least squares estimation and backpropagation for membership function parameter estimation [44].

Several fuzzy inference systems are described by different researchers [45]. The most commonly-used systems are the Mamdani-type and Takagi–Sugeno type, also known as Takagi–Sugeno–Kang type. In the case of Mamdani-type fuzzy inference system, both premise (if) and consequent (then) parts of a fuzzy if-then rule are fuzzy propositions. In the case of a Takagi–Sugeno-type fuzzy inference system where the premise part of a fuzzy rule is a fuzzy proposition, the consequent part is a mathematical function, usually a zero- or first-degree polynomial function [45, 47].

The advantages of FL for grade estimation is clear because it proposes a powerful tool that is flexible and in lack of data with its ability which is if-then rules would able to solve the problems [49]. As discussed, one of the biggest problems in FL application is the shape and location of membership function for each fuzzy variable which solves by trial and error method only. In contrast, numerical computation and learning are the advantages of NN, however, it is not easy to obtain the optimal structure (number of hidden layer and number of neuron in each hidden layer, momentum rate and size) of constructed NN and also this kind of artificial intelligent is more based on numerical computation rather that than symbolic computation.

Both FL and NN have their advantages, therefore, it is good idea to combine their ability and make an strong tool and also a tool which improve their weak as well as lead to least error. Jang [44, 50] combined both FL and NN to produce a powerful processing tool named NFSs which is a powerful tool that have both NN and FL advantages and the most common one is ANFIS.

Assume that the considered FIS has two inputs x and y and one output f (Fig. 8.9). A first-order Sugeno fuzzy model, a common rule set with two fuzzy if–then rules proposed by Jang [50]:

Rule 1: If x is A_1 and y is B_1 , then $f_1 = p_1 x + q_1 y + r_1$ (8.14)

Rule 2: If x is A_2 and y is B_2 , then $f_2 = p_2 x + q_2 y + r_2$ (8.15)

Fig. 8.10 illustrates the reasoning mechanism for the Sugeno model [46]. The corresponding equivalent ANFIS. This output W_1 and W_2 is the input into layer 4 of Fig. 8.10:

$$f_1 = p_1 x + q_1 y + r_1 \qquad (8.16)$$

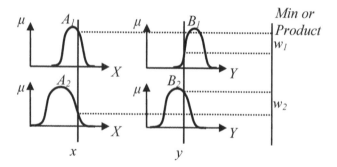

Fig. 8.9 First-order Sugeno fuzzy model

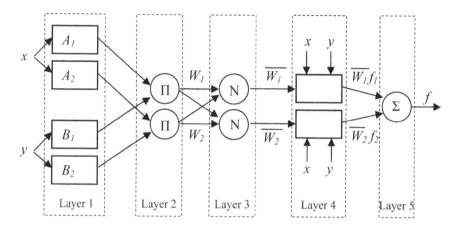

Fig. 8.10 Equivalent ANFIS architecture with two inputs and one output

$$f_2 = p_2x + q_2y + r_{21} \qquad (8.17)$$

$$f = \frac{W_1 f_1 - W_2 f_2}{W_1 - W_2} = \overline{W}_1 f_1 - \overline{W}_2 f_2, \qquad (8.18)$$

Layer l.

$O_{l,i}$ is the output of the i-th node of the layer l. Every node i in this layer is an adaptive node with a node function:

$$O_{1,i}=\mu A_i(x) \text{ for } i=1, 2, \text{ or } O_{1,i}=\mu B_{i-2}(x) \text{ for } i=3, 4, \qquad (8.19), (8.20)$$

where x (or y) is the input node i and A_i (or B_{i-2}).

Therefore $O_{1,i}$ is the membership grade of a fuzzy set (A_1, A_2, B_1, B_2). Typical membership function:

$$\mu A(x) = \frac{1}{1 + \left| \frac{x - c_i}{a_i} \right| 2b_i}, \tag{8.21}$$

where a_i, b_i, c_i are the parameter set.

Parameters are referred to as premise parameters.

Layer 2.

Every node in this layer is fixed node. The output is the product of all the incoming signals according to

$$O_{2,i} = W_i = \mu A_i(x) \bullet \mu B_i(y), \quad i=1, 2. \tag{8.23}$$

Each node represents the fire strength of the rule. Any other T-norm operator that perform the AND operator can be used

Layer 3.

Every node in this layer is fixed node with function of normalization labelled N. The i node calculates the ratio of the i rulet's firing strength to the sum of all rulet's firing strengths.

$$O_{3,i} = \overline{W}_i = \frac{W_i}{W_1 + W_2} \quad i=1, 2. \tag{8.24}$$

Outputs are called normalized firing strengths.

Layer 4.

Every node i in this layer is adaptive node with a node function:

$$O_{4,1} = \overline{W}_i f_i = \overline{W}_i (p_x + q_i p + r_i), \tag{8.25}$$

where $\overline{W}_i$ is the normalized firing strength from layer 3, (p_i, q_i, r_i) is the parameter set of this node. These are referred to as consequent parameters.

Layer 5.

The single node in this layer is a fixed, which computes the overall output as the summation of all incoming signals:

$$overall\ output = O_{5,1} = \sum_i \overline{W}_i f_i = \frac{\sum_i \overline{W}_i f_i}{\sum_i W_i} . \qquad (8.26)$$

8.3. Neural network method for identification of the multisensor's individual characteristic curve

The basic idea of the proposed method [18, 43] is that the increase of the accuracy identification of the MS ICC is possible by increasing the amount of information for training of approximate NN. This is provided by using the calibration results of group one-type MS in such number of points that it's enough to identify the MS ICC this type with a specified accuracy. It allows the NN to reveal and generalize patterns of relationships which are characteristic for this MS characteristic curve. According to these identified characteristic patterns of relationships and the calibration results this MS for reduced amount of points, the calibration results for other points are predicted. It allows getting the calibration results of this MS for the complete set of points – those for which the calibration has been really performed as well as for other points of characteristic curve for which the calibration has not been really performed.

The basis for obtaining the high-precision prediction is a database of the calibration group one-type MS. This database has collected in the amount of points which are enough to identify MS ICC this type. This characteristic curve need to hold with a specified accuracy, as well as individual NN prediction at each point according to methods described below. Using a set of NN, each of which specializes in the prediction results of calibrating at one particular point, on the basis of calibration results at other points that every time belongs to the strictly specified set, allows to increase the accuracy of prediction. For example, each NN uses the calibration results that are collinear with the point which calibration value is predicted. This straight line is specified as common line: for points of real calibration given MS, for calibration points of the one-type MS and for the point which calibration result is predicted. Thus, it's achieved one-one correspondence conditions of NN training and prediction. It contributes to high precision training of NN and as a result the high accuracy of prediction.

Implementation of the proposed method is illustrated by example of two-parameter sensor. This is the abstracted example which demonstrates the capability of the proposed method in general. At the first step, it's necessary to carry out a real calibration for group of 30 ... 50 one-type MS in a large amount of points, for example, in 49 points (in 7 calibration points for each of two physical quantities). The location of calibration points in the conversion ranges of both physical quantities, under the condition its uniform distribution, is shown on Fig. 8.11. The values of a first physical quantity X_1 are laid out on the axis of abscissas, the values of the second physical quantity X_2 are laid out on the axis of ordinates. The 49 calibration points of the one-type MS, in the points 11 ... 77, are shown by thin rings.

The MS, which ICC is defined, has been calibrated at 9 points, marked on Fig. 8.11 by bold rings. They are points with the following numbers 22, 24, 26, 42, 44, 46, 62, 64, 66. The first digit in a point number corresponds to the point of physical quantity X_1, and the second one corresponds to physical quantity X_2. These calibration points correspond to the calibration points of one-type MS, however there are the 9 points only.

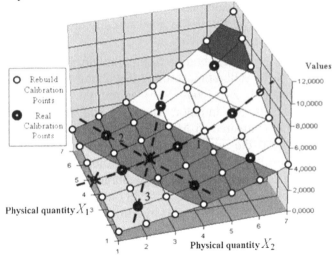

Fig. 8.11 ICC of MS and placement of real calibration points and recoverable calibration points.

A task of the proposed method is the prediction of the value calibration results at all calibration points (thin rings) on the basis of the MS real calibration at 9 points (bold rings). For solving this task can be used the calibration results of one-type MS (thin rings). Criteria of quality of the task execution is a high precision of predicted results for the calibration in points: 11, 12, - 16, 17, 21, 23, 25, 27, 31, 32, - 36, 37, 41, 43, 45, 47, 51, 52, - 56, 57, 61, 63, 65, 67, 71, 72, - 76, 77.

On the first stage, the MS calibration results are predicted, choosing its points with real calibration, and points of calibration one-type MS that are collinear with the predicted point of MS characteristic curve, which is calibrated. This process is illustrated on Fig. 8.11 for predicted point 34 of MS characteristic curve. It is possible to draw three straight lines through the point 34. These lines contain the points of MS real calibration. The straight line #1 passes through the points 24, 44, 64; the straight line #2 through the points 26, 42; the straight line #3 through the points 22, 46. Simultaneously the straight line #1 includes the following calibration point of the one-type MS 14, 34, 54, and 74. Each line corresponds to its own NN.

The straight line #1 has got a maximum number of coincidence points for points of MS calibration and the group of one-type MS. Let's consider the process of training NN on its example. The training sample of this NN is formed as follows. The one MS from the group, for example, number 1 is adopted as a calibrating MS called imitator. Others MS are the group of one-type MS which were calibrated in a large number of points Therefore, the first training vector is formed as a set of calibration results, which are placed, for example, in the following order:

(i)   The values of the MS calibration results-imitator in points 24, 44, 64.

(ii)   The values of the calibration results MS (among of the group one-type MS in points 14, 24, - 64, 74) that are the most similar to the calibration results MS-imitator. The most similar MS (among group one-type MS) compared to MS, adopted as the MS-imitator, is one, which has got the minimal sum absolute values of deviations calibration results in that points of real calibration, which belongs to straight line, according to which, the calibration values is predicted.

(iii) The values of the calibration results MS (among a group of similar MS in points 14, 24, - 64, 74) the next by similarity to MS-imitator.

(iv)   The values of the MS calibration results are placed in decreasing order of the similarity to MS-imitator.

(v)   The values of the calibration results MS (among a group of similar MS in points 14, 24, - 64, 74) which has the least similarity to MS-imitator.

(vi)   The values of the calibration results MS (adopted as the MS-imitator) at point 34.

Thus, by sorting the calibration results according to (i)-(vi), the MS's (in the structure of the training vector) are placed in decreasing order of their similarity to MS-imitator, precisely in those points where a prediction of the calibration values is running.

The last element of training vector will be a value which is compared with the result of the NN operation. It should be noted that the order of recording values (i)-(vi) for calibration results must be always the same. In such case the NN during training is able to detect patterns of relationships that determine the deviation of MS characteristic curve of this type from the nominal one. Therefore, the order itself is not important, constancy (invariability) is important for training as well as for prediction.

The next vector of training sample is formed by identifying other MS-imitator (among the group of one-type MS which was calibrated at all points of calibration). To form the training sample, the items (i)-(vi) should repeat again. The process formation of training vectors continues until all MS of group one-type MS (which really have been calibrated in all points of calibration) perform the role of MS-imitator.

Each vector of the training set has the following structure N_{24}^0, N_{44}^0,

$$N_{64}^0, \ N_{14}^{+1}, \ N_{24}^{+1}, \ N_{34}^{+1}, \ N_{44}^{+1}, \ N_{54}^{+1}, \ N_{64}^{+1}, \ N_{74}^{+1}, \ N_{14}^{-1}, \ N_{24}^{-1}, \ N_{34}^{-1}, \ N_{44}^{-1},$$

$$N_{54}^{-1}, \ N_{64}^{-1}, \ N_{74}^{-1}, \ N_{14}^{+2}, \ N_{24}^{+2}, \ ..., \ N_{14}^{-2}, \ N_{24}^{-2}, \ ..., \ N_{14}^{+n}, \ N_{24}^{+n}, \ ..., \ N_{14}^{-n},$$

$$N_{24}^{-n}, \ ..., \ N_{64}^{-n}, \ N_{74}^{-n}.$$

The lower indexes above indicate the calibration points. The upper indexes are increasing with the raising of the absolute difference between the MS calibration results with index 0 and N. The signs "+" and "-" indicate the polarity of the difference. At predicting the N_{24}^0, N_{44}^0, N_{64}^0 there are used the calibration results of MS which ICC is identifying.

After training of the NN with this training sample its coefficients express the regularity of deviation the MS characteristic curve this type from the nominal one. This trained NN can be used to predict the results of MS -imitator calibration in point 34 accordingly to the straight line #1.

It's recommended to use the trained NN for predicting the calibration results of MS which is calibrating. Then results of its calibration in points 24, 44, 64 are putting in the prediction vector: in the position of placing the calibration results of MS-imitator during NN training at the same points according to (i) above. After that the calibration results for the most similar MS are placed in the prediction vector, then the next one (according to (ii), (iii), (iv)). The values of results calibration for the least similar MS are rejected. As a result the NN predicts with high accuracy the values of MS calibration result at point 34 thru the line # 1.

For predicting calibration results of MS-imitator, at point 34, according to lines #2 and #3, should repeat all described operations again and train the appropriate NN. For predicting calibration results in other points the appropriate NN should be trained. It should take into account also the similarity of the way for drawing direct lines through points 43, 45, 54 to lines through point 34. Thru points 33, 35, 53, 55 it is possible to draw two straights through points 22, 44, 66 and through the points 24, 42.

Thru points 11, 12, 16, 17, 21, 23, 25, 27, 31, 32, 36, 37, 41, 47, 51, 52, 56, 57, 61, 63, 65, 67, 71, 72, - 76, 77 it is possible to draw only one line.

For prediction can be used the three-layer perceptron [17] with the number of input neurons that correspond to the number of the calibration results at given point. The output neuron is one. The amount of hidden layer neurons and the type of activation function should be chosen experimentally by results of trial training. To ensure good generalizing properties the number of neurons in the hidden layer should be selected as a minimal one according to [39]. As a result the reduction of calibration points in about 80% because a real ICC calibration is carried out in 9 of 49 points only (Fig. 8.11).

To assess the accuracy of the proposed method it can carry out a research by a simulation modeling method. For this purposes initially was developed an appropriate model of errors MS. Error of ICC identification method the nominal conversion curve can be described by polynomials [51]:

$$Y_{NOM} = (A(X_1 + B)^k + C(X_1 + B)) \bullet (D(X_2 + E)^l + F(X_2 + E)) \bullet G, \quad (8.27)$$

where X_1, X_2 are measurable physical quantities, for example photoelectric current and temperature respectively; $A...G$, k, l are coefficients and indexes of polynomials respectively; Y_{NOM} is MS nominal output.

Obviously, the MS additive and multiplicative errors can be corrected without using the NN. Therefore, it should be useful to study the influence of the nonlinear component of MS error on the correction result. The MS errors of different physical quantities can be distinct not only quantitatively but also qualitatively. There were explored various combinations of errors model described by function:

$$Y = Y_{NOM} \pm n\Delta\left(\pm K_1(i-4)^a \pm K_2(j-4)^b\right), \quad (8.28)$$

where n is the number of combinations, it was taken $n = 100$ (i.e. 50 experiments for each polarity); Δ – quantization step of MS error, taken 0.1% (i.e. maximum error MS for each physical quantity is equal 5%); K_1, K_2 are the coefficients that characterize the nonlinearity of MS error function, and equal 1%; exponents a and b can be equal 2, 3, 4 and higher orders.

Furthermore the output signal of MS is distorted by random error (random noise):

$$Y_N = Y + Rnd(K_3), \qquad (8.29)$$

where K_3 is the coefficient determining the amplitude of random error.

8.4. Improved approach for identification of the multi-sensor's individual characteristic curve

Proposed approach [23] (Fig. 8.12) includes three new procedures: formation of training and predicting vectors for ANFIS; ANFIS training; prediction using ANFIS.

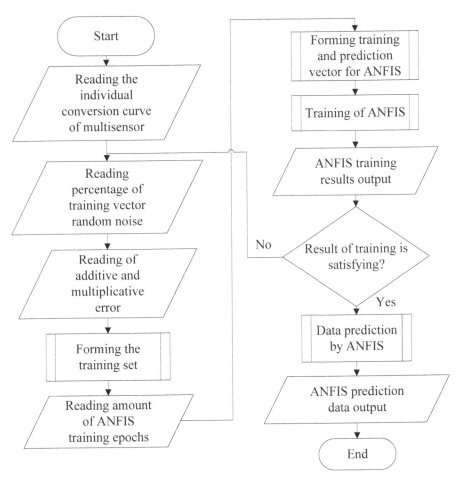

Fig. 8.12 Algorithm of intelligent data processing of MS based on ANFIS

The procedure formation of training and predicting vectors for ANFIS consists of the following steps: the main module UAITF (Fig. 8.13) which forming the training programs and predictor vectors for ICC, ANFIS reads data of MS, reads the percentage of random error for training, writes model of additive and multiplicative errors and their values (degree polynomials and their signs according to [18, 43]). Next performed the procedure for forming the training sample: hosted actual calibration point values that are straight, placed the value of the whole line, the following value is placed next to similarity, next placed the value of the least similar (most distant), recently hosted a value that is the result, the purpose of calibration. After reading the sorted data read periods of training and called module UTest (Fig. 8.13), which provides interaction of GUI programs with Matlab programs. Module UTest scanned data in the appropriate format. The module also records start time and calculate duration, it transfers those values into the module UAITF for display. Then control is switched to the module CreateAndTrainANFIS (Fig. 8.13), which generates a FIS-structure and framework of rules, reads the settings for training and initial values, train and receive the results.

After completing training of ANFIS, the user can make data prediction. This module named Approximation, is used (Fig. 8.13), with the following sequence: first, through a module UTest received input vector for the approximation, – goal of prediction (last column).

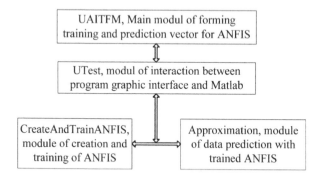

Fig. 8.13 Generalized structure modules

Then using readfis – toolkit of Fuzzy Logic Toolbox, package of Matlab, Fis-loaded structure from a saved file from trained network. Then, the approximation of objective prediction by processing input vector using the trained network performance and function of fuzzy inference with the command evalfis. Result of approximation using the module UTest transferred into the main module UAITF for mapping and charting goals transferred into the main module UAITF for mapping and charting goals prognosis and prediction of the result. The main modules of the results can be saved as an image file.

8.5. Experimental studies

The main graphical interface was implemented in the environment Borland C ++ Builder, and software in the controller.

Experimental studies of the controller shown that increasing the number of training epochs system reduces error and increases the accuracy of prediction data. At 2.5% of random error in training vectors, where the number of training epochs over 1000, the system is overtrained. Therefore, it is most effectively using the number of epochs for training, less than 1000.

In Table 8.1 and Fig. 8.14 the results of ANFIS training are put, which was trained at 2.5% of random error in training vectors, the number of training epochs, equal 20, 200, 400, and 1000.

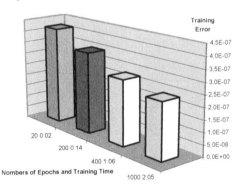

Fig. 8.14. Chart of ANFIS training errors at 2.5% of random error, depending on the number of epochs and training time

Table 8.1. Training the ANFIS with training vector random error 2.5%

Experiment number	Number of epochs	Training Time (hr: min.)	Error of training
1	20	0:02	4.2E-07
2	200	0:14	3.5E-07
3	400	1:06	2.8E-07
4	1000	2:05	2.4E-07

There are the following advantages of the proposed correspond algorithm for the intelligent data processing of MS: accelerating training and predicting stages, which is achieved by single study and a small number of training epochs; value stability of training results (it doesn't dependent on runs number); the ability of the suppressing random error of large amplitude.

Thus, based on the integration of the modified method of identifying the ICC of MS and the adaptive neuro-fuzzy inference system algorithm with intelligent processing of MS signals, which combines the advantages of NN and FL, which helped to reduce the number of training epochs and time of network training correspondingly.

8.6. Experimental results

According to the considered method it has been researched the errors of prediction of calibration results of MS at point 34 (see fig. 8.11) by using ANFIS. In Figs. 8.15 and 8.16 are shown typical graphs of prediction errors from a particular variant of test sample, which does not coincide with the training set. Figure 8.15 shows the prediction error of the random error 2.5%, for the exponent $a= +2$ and $b= +2$ described by (8.28).

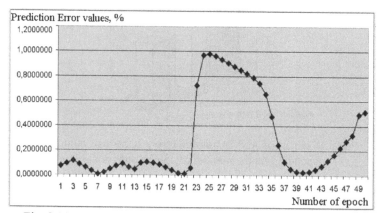

Fig. 8.15 Error values for formula (8.28), with exponents $a= +2$ and $b= +2$ at random error 2.5%

Figure 8.16 shows the prediction error of the random error 2.5%, for the exponent $a= -4$ and $b= -2$ according to (8.28). As can be seen from the graphs, the implementation of the identification method of ICC of MS by ANFIS is similar to the implementation of the method based on NN [43, 18]. It has property to suppress random error.

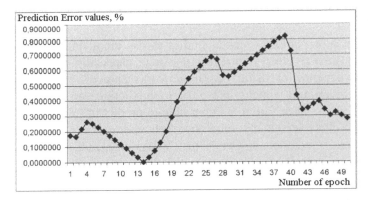

Fig. 8.16 Error values for formula (8.28), with exponents $a = -4$
and $b = -2$ at random error 2.5%

Figure 8.17 shows the dependence of the suppression coefficient from amplitude of random error for NN method. Figure 8.18 shows the dependence of the suppression coefficient from amplitude of random error for ANFIS. The results of detailed research of prediction errors for realization of identification method of ICC of MS with using ANFIS is presented in Table 8.2. In this table the top number represents the maximum value of prediction error, the lower numbers represents the mean error according to selected values of "a" and "b".

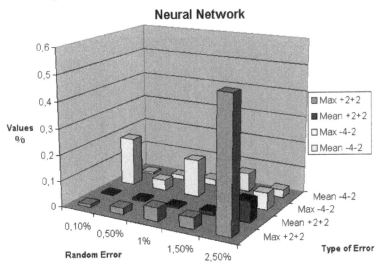

Fig. 8.17 Relationship of the suppression coefficient from amplitude of random error for neural network method

8.7. Conclusion

As where shown in experimental studies above, the main advantage of the ANFIS is the high effect on the training stage, especially for small training times. Experience of using neural networks showed that in the small amounts of training epochs, the error of prediction can vary by several tenths of times. Therefore it is impossible to predict this error. The ANFIS system is characterized by otherwise behaviour. The resulting error of training and prediction stages is just weakly depends on the launches sequence and training time (quantity of training epochs). As it's shown in Table 8.1, the error is decreased less than twofold during increasing the training time by 50 times. However, even at short training times we got rather a small error. This makes the ANFIS a promising system for implementation in real-time applications, and primary advantages of ANFIS will be used there. The identified disadvantage of ANFIS is a little bit bigger training error in comparison with neural networks (see Table. 8.2).

One of the perspective research in future is investigating the effectiveness of proposed algorithm with the variety of the training epochs range, training time, and etc.

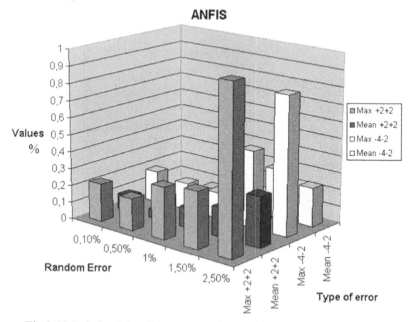

Fig 8.18 Relationship of the suppression coefficient from amplitude of random error for ANFIS

Table 8.2. Maximum relative / mean errors of data prediction vs percentage of random error for neural network and ANFIS according to formula (8.28)

Percentage of Random Error, %	Maximum Relative / Mean Errors, %			
	Neural Network (Perceptron)		ANFIS	
	a=+2, b=+2	a=−4, b=−2	a=+2, b=+2	a=−4, b=−2
0.1	0.011 / 0.004	0.186 / 0.006	0.222 / 0.092	0.187 / 0.018
0.5	0.025 / 0.008	0.041 / 0.016	0.184 / 0.050	0.163 / 0.075
1	0.052 / 0.007	0.143 / 0.023	0.305 / 0.119	0.153 / 0.333
1.5	0.043 / 0.015	0.153 / 0.070	0.333 / 0.170	0.453 / 0.291
2.5	0.502 / 0.082	0.065 / 0.030	0.981 / 0.294	0.819 / 0.228

References

[1] Taner A. and Brignell J. Virtual instrumentation and intelligent sensors. *Sensors and Actuators A: Physical*, Vol. 61, No. 1-3. 1997, pp. 427-430.

[2] http://www.figarosensor.com/products/813pdf.pdf

[3] Capone S., Siciliano P., Barsan N., Weimar U., and Vasanelli L. Analysis of CO and CH4 gas mixtures by using a micromachined sensor array. *Sensors and Actuators B: Chemical*, Vol. 78, No. 1-3, 2001, pp. 40-48.

[4] Martin M., Santos J., and Agapito J. Application of artificial neural networks to calculate the partial gas concentrations in a mixture. *Sensors and Actuators B*, Vol. 77, No. 1-2, 2001, pp. 468-471.

[5] Srivastava A. Detection of volatile organic compounds (VOCs) using SnO2 gas-sensor array and artificial neural network. *Sensors and Actuators B*, Vol. 96, No. 1-2, 2003, pp. 24-37.

[6] Patent number 50830 Ukraine, G06F15/18. "The method of neural network training samples generation of drift prediction device for data acquisition", A. Sachenko, V. Kochan, V. Turchenko, (Ukraine), V.Golovko, U. Savitskiy (Belarus), T. Laopoulos (Greece), declared 04.01.2000, printed 15.11.2002, 14p. (In Ukrainian).

[7] Brignell E. Digital compensation of sensors. *Scientific Instruments*, Vol. 20, N 9, 1987, pp. 1097-1102.

[8] International standard IEC 751.

[9] Rogelberg I. *Alloys for termocouples*. Moscow: Metallurgy, 1983, 360p. (In Russian).

[10] Daqi G., Shuyan W., and Yan J. An electronic nose and modular radial basis function network classifiers for recognizing multiple fragrant materials. *Sensors and Actuators B*, Vol. 97, No. 2-3, 2004, pp. 391-401.

[11] Zhang H., Balaban M., and Principe J. Improving pattern recognition of electronic nose data with time-delay neural networks. *Sensors and Actuators B*, Vol. 96, No. 1-2, 2003, pp. 385-389.

[12] Llobet E., Brezmes J., Gualdron O., Vilanova X., and Correig X. Building parsimonious fuzzy ARTMAP models by variable selection with a cascaded

genetic algorithm: application to multisensor systems for gas analysis, *Sensors and Actuators B*, Vol. 99, No. 2-3, 2004, pp. 267-272.

[13] Gijbels I. and Verhasselt A. *P-splines regression smoothing and difference type of penalty*. Springer Science + Business Media, 2009, 499p.

[14] Ortega A., Marco S., Perera A., Sundic T., Pardo A., and Samitier J. An intelligent detector based on temperature modulation of a gas sensor with a digital signal processor. *Sensors and Actuators B*, Vol. 78, No. 1-3, 2001, pp. 32-39.

[15] Turchenko I., Osolinsky O., Kochan V., Sachenko A., Tkachenko R., Svyatnyy V. and Komar M. Approach to neural-based identification of multisensor conversion characteristic. In *5th IEEE International Workshop on Intelligent Data Acquisition and Advanced Computing Systems: IDAACS'09*, Rende, Italy, 2009, pp. 27-31.

[16] Turchenko I. and Kochan V. Improvement of identification accuracy of multisensor conversion characteristic using SVM. In *6th IEEE International Workshop on Intelligent Data Acquisition and Advanced Computing Systems: IDAACS'11*, Prague, Czech Republic, 2011, pp. 388-392.

[17] Kroese B. *An Introduction to Neural Networks*. Amsterdam: University of Amsterdam, 1996, 120p.

[18] Roshchupkin O., Smid R., Kochan V., and Sachenko A. Reducing the calibration points of multisensors. In *Proceedings of the 9th IEEE International Multi-Conference on Systems, Signals and Devices (SSD'2012)*, Chemnitz, Germany, March 20-23, 2012, pp.1-6.

[19] Depari A., Flammini A., Marioli D., and Taroni A. Application of an ANFIS algorithm to sensor data processing . In *Proc. of the IEEE Instrumentation and Measurement Technology Conference, 2005. IMTC 2005*, Ottawa, Canada, Vol. 3, 17-19 May 2005, pp. 1596-1599.

[20] Auge J., Dierks K., Eichelbaum F., and Hauptmann P. High-speed multi-parameter data acquisition and web-based remote access to resonant sensors and sensor arrays. *Sensors and Actuators B*, Vol. 95, No. 1-3, 2003, pp. 32-38.

[21] Michie D., Spiegelhalter D.J., and Taylor C.C. *Machine Training, Neural and Statistical Classification*. New York: Ellis Horwood, 1994. 290 pp.

[22] Isermann R. *Fault-Diagnosis Systems*. Springer, Berlin Heidelberg, 2006. 475p.

[23] Isermann R. and Balle P. Trends in the application of model-based fault detection and diagnosis of technical processes. *Control Engineering Practice*, Vol. 5 (5), 1997, pp. 709-719.

[24] Laprie J.C. On computer system dependability and un-dependability: Faults, errors, and failures. IFIP WG 10.4, Como, Italy, 1983.

[25] Fraden J. *Handbook of Modern Sensors: Physics, Designs, and Applications*. Springer, 2010, 663 pp.

[26] Bakiotis C., Raymond J., and Rault A. Parameter and discriminant analysis for jet engine mechanical state diagnosis. In Proc. of The 1979 IEEE Conf on Decision & Control, Fort Lauderdale, USA, 1979.

[27] Isermann R. Parameter-adaptive control algorithms – A tutorial. *Automatica*, Vol. 18(5). 1982, pp. 513-528.

[28] Isermann R. Process fault detection on modeling and estimation methods – A survey. *Automatica*, Vol. 20(4), 1984, pp. 387-404.

[29] Filbert D. and Metzger, L. Quality test of systems by parameter estimation. In *Proc. 9th IMEKO-Congress*, Berlin, Germany, May 1982.

[30] Filbert D. Fault diagnosis in nonlinear electromechanical systems by continuous-time parameter estimation. *ISA Trans.*, Vol. 24(3), 1985, pp. 23-27.

[31] McCulloch W. and Pitts W. A logical calculus of the ideas immanent in nervous activity. *Bull. Math. Biophys.*, Vol. 5, 1943, pp. 115-133.

[32] Rosenblatt F. The perceptron: a probabilistic model for information storage &organization in the brain. *Psychological Review*, Vol. 65, 1958, pp. 386-408.

[33] Widrow B. and Hoff M. Adaptive switching circuits. IRE WESCON Conv. Ree., 1960, pp. 96-104.

[34] Hecht-Nielson R. *Neurocomputing*. Addison-Wesley, Reading, 1990, 433 pp.

[35] Haykin S. *Neural networks*. Macmillan College Publishing Company, Inc., New York, 1994, 696 pp.

[36] Bishop C. *Neural networks for pattern recognition*. Oxford University Press, Oxford, 1995, 482 pp.

[37] Jain A.K., Mao J., and Mohiuddin K.M. Artificial neural networks: A tutorial. *Computer*, March 1996, pp. 31-44.

[38] Golovko V.A. *Neirointelligence: Theory and Application. Book #1: Organization and Training of the Neural Networks with Forward and Backward Connections*. Brest. Publisher BTI, 1999, 264 pp. [in Russian].

[39] Sachenko A., Kochan V., Turchenko V., Golovko V., Savitsky J., Dunets A., and Laopoulos T. Sensor errors prediction using neural networks. In *Proceedings of the IEEE-INNS-ENNS International Joint Conference on Neural Networks (IJCNN'2000)*, Como (Italy). Vol. IV, 2000, pp. 441-446.

[40] Rumelhart D., Hinton G., and Williams R. Learning representation by backpropagation errors. *Nature*, Vol. 323, 1986, pp. 533-536.

[41] Montgomery D.C. Forecasting and time series analysis. In D.C. Montgomery, L.A. Johnson, J.S. Gardiner, 2nd ed., 381 pp.

[42] Patent of Ukraine for Inventions # 103802 registered on 25.11.2013. Application # a2011 13840, declared 24.11.2011. Method of Multisensor Convertion Function Identification. Roshchupkin O. U., Kochan V. V., Sachenko A. O. [in Ukrainian].

[43] Jang J.-S.R. ANFIS: Adaptive-network-based fuzzy inference systems. *IEEE Transactions on Systems, Man, and Cybernetics*, Vol. 23, No. 3, 1993, pp. 665-685.

[44] Mamdani E.H. Applications of fuzzy algorithm for control of a simple dynamic plant. *Proc. IEEE*, Vol. 121 (12), 1974, pp. 1585-1588.

[45] Sugeno M. and K. Tanaka. Successive identification of a fuzzy model and its application to prediction of complex systems. *Fuzzy Sets Syst.*, Vol. 42. 1991, pp. 315-334.

[46] Takagi T. and M. Sugeno. Fuzzy identification of systems and its applications to modeling and control. *IEEE Trans. Syst., Man Cybern.*, Vol. 15 (1), 1985, pp. 116-132.

[47] Zadeh L.A. Fuzzy sets. *Information and Cont.*, Vol. 8, 1965, pp. 338-353.

[48] Pejman T. and Ardeshir H., Application of adaptive neuro-fuzzy inference system for grade estimation; Case study, Sarcheshmeh Porphyry Copper Deposit, Kerman, Iran. *Australian Journal of Basic and Applied Sciences*, Vol. 4(3), 2010, pp. 408-420.

[49] Jang J.S.R. Neuro-fuzzy modeling: architecture, analyses and applications, Dissertation, Department of Electrical Engineering and Computer Science, University of California, Berkeley, CA 94720, 1992.

[50] Turchenko I., Kochan V., and Sachenko A. Accurate recognition of multi-sensor output signal using modular neural networks. *International Journal of Information Technology and Intelligent Computing*, Vol. 2(1), 2007, pp. 27-47.

Biographies

Nataliya Roshchupkina received her master's degree at Computer Systems and Networks Department, Chernivtsi National University, Ukraine and she has worked here as an assistant professors at the Department of Computer Systems and Networks since 2004. She is currently working in her PhD at the same department. Her research interests include Artificial Intelligence, Measuring Systems, Multisensor Systems, Neural Networks, Adaptive Neural –Fuzzy Inference Systems.

Volodymyr Kochan received Ph. D. degree in 1989. From 1996 he was Assistant Prof. at Department of Information-Computing System and Control and since 2004 Director of the Research Institute of Intelligent Computer Systems. His research interests are focused to Intelligent Measurement Devices and Measurement Systems.

Anatoliy Sachenko received his PhD Degree in Electrical Engineering from L'viv Physics and Mechanics Institute, Ukraine in 1978, and his Doctor of Technical Sciences Degree in Electrical and Computer Engineering from Leningrad Electrotechnical Institute, Russia in1988. He is currently Professor and Head of the Information Computer Systems and Control Department at the Ternopil National Economic University, and a Principal Investigator of Research Institute for Intelligent Computer

Systems, and he is a part-time Professor of Organization and Management Department at the Silesian University of Technology, Poland, and he is an Affiliated Professor of the Electrical and Computer Engineering Department at the University of New Hampshire. He has been the Honored Inventor of Ukraine since 1991, the IEEE Senior Member since 1993, the IEEE TNEU Student Branch Counselor since 1999, a Chair of I&M / Computational Intelligence Joint Societies Chapter, IEEE Ukraine since 2005. His research interests include Artificial Neural Network Application, Distributed Sensor Networks, Image Processing and Pattern Recognition, Computation Intelligence for Homeland Security and Safety, High Performance Computing and Grids, Wireless Sensor Networks.

Oleksiy Roshchupkin. acquired Master's Degree in 2004 form Chernivtsi National University, Ukraine. He has been at Department of Computer Systems and Networks at the same university since 2005. He is currently working in his Ph.D. at the Faculty of Electrical Engineering, at Czech Technical University in Prague. His main research activities focus on Information-Measuring Systems, Microcontrollers, Multisensor Systems, Neural Networks, Sensors.

Radislav Smid received his Ph.D. degree in Measurement and Instrumentation from the Czech Technical University in Prague in 2000. He is currently head of Laboratory for Diagnostics and Non-destructive Testing and an associate professor at the Department of Measurement of Faculty of Electrical Engineering, Czech Technical University in Prague. His major research interests include Signal Processing and Analysis for Diagnostics, Biomedical Apps and Communications. Non-destructive Testing (NDT), Machine Condition Monitoring (MCM), Automated Testing, Fault Detection and Diagnosis.

9

FPGA-based ANFIS linearizer for measurement systems

Mounir Bouhedda

*Laboratory of Advanced Electronic Systems (LSEA),
Department of Electrical Engineering and Computers,
Faculty of Sciences and Technology, University of Medea, Medea, Algeria*

Abstract

Nonlinear sensors and digital solutions are used in many data acquisition system designs. As the input/output characteristic of most sensors is nonlinear in nature, obtaining data from a nonlinear sensor by using an optimized device has always been a design challenge. This chapter aims to present a new Adaptive Neuro-Fuzzy Inference System (ANFIS) digital architecture based Field Programmable Gate Array (FPGA) to linearize the sensor's characteristic.

Keywords: data acquisition system, FPGA, linearization, neuro-fuzzy, sensor

9.1 Introduction

Sensors are fundamental elements in many measurement circuits. They are used in many industrial applications related to instrumentation and control. Sensors generally, take a certain form of physical magnitude (temperature, pressure, etc.) and convert it to analog output. This latter incorporate an electronic circuit in order to present a suitable characteristics for the next stages of the measurement chain [1]. However, most of sensors have nonlinear characteristic in nature from which a linear output is desired.

Sensor linearization in digital environment utilizing the entire range of the characteristic is a crucial step in the instrument conditioning process. Several linearization techniques have appeared in many research works. These techniques can be organized into three main classes. Analog

V. Haasz and K. Madani (Eds.), Advanced Data Acquisition and Intelligent Data Processing, 213–231.

hardware-based linearization circuit [2, 3], software based linearization algorithms [4–8] and hybrid analog-digital solutions [9].

Analog circuits are frequently used for improving the linearity of sensor characteristics, which implies additional analog hardware and typical problems particular to analog circuit such as temperature drift, gain and offset errors. Using the second technique, sensor nonlinearities can be compensated by means of arithmetic operations, if an accurate sensor model is available (direct computation of the polynomials), otherwise with the use of multidimensional look-up tables. Direct computation of the polynomial method is more accurate but requires a longer time for computation, while the look-up table method, though faster, is not very accurate [6–10]. The third technique is performed by interfacing a passive or an active nonlinear analog circuit between the sensor and an analog to digital converter (ADC) [9].

Obtaining an accurate mathematical model for a system with tacking account disturbing parameters can be rather complex and time consuming. This fact has led the researches to operate the neural and fuzzy techniques in modelling complex systems utilizing the input-output data sets. Although fuzzy logic allows systems modeling using experience and human knowledge with if–then rules, it is not always easy to transform a human knowledge into the rule base of a fuzzy inference system. In addition, fuzzy parameters have to be turned effectively. On the other hand, artificial neural networks (ANNs) can be advantageous because they provide an adjusting mechanism and hence they are used to optimize the fuzzy system. The combination of ANNs with Fuzzy systems create a new robust model called Adaptive Neural Fuzzy Inference System (ANFIS) [11].

Fuzzy set theory has been applied to a wide range of problems in electronic instrumentation, automatic control, signal treatment, data clustering, pattern recognition, and data classification. Recently, application of neural networks and fuzzy logic techniques has emerged as a promising area of research in the field of instrumentation and measurement [12–15]. Engin et al. in [16] demonstrate the modeling capability of neuro-fuzzy systems. It is shown that the ANFIS can modelize a nonlinear system very accurately by means of given data [11].

ANFIS which is a specific approach in neuro-fuzzy development, has shown significant promise in modelling nonlinear functions. It learns features of the data set and adjusts the fuzzy system characteristics according to a given error criterion.

FPGAs belong to the wide family of programmable logic component [17], their densities are now exceeding 10 million gates [18]. FPGAs can be defined as a matrix of configurable logic blocks (combinatorial and/or sequential), linked to each other's by an interconnection network that is also entirely reprogrammable. FPGAs technology allows developing specific hardware architectures within a flexible programmable environment. This specificity of FPGA gives the designer a new degree of freedom comparing to microprocessors implementation, since the hardware architecture of the synthesized system is not imposed a priori.

This chapter is organized as follows: after an introduction in section one. In section two, a brief presentation of fuzzy systems is given with focusing on ANFIS approach. In section three, the different steps of neuro-fuzzy modelization sensor linearizer is presented, section four is devoted to FPGA implementation where a digital description is given; for characterization and comparative purposes, the performance of the developed architecture is examined with comparison to ANFIS software model and with two other architectures based FPGA using classical techniques in section four. Finally, the conclusions are pointed out in section five.

9.2 The linearizer in a measurement system

An acquisition chain collects the necessary information and the knowledge to control a process; it delivers this information in a digital form suitable to their exploitation [19, 20].

An industrial process contains many elements whose physical interactions contribute to the aim of manufacturing or processing; the process is instrumented and controlled for this purpose. The state of a process is at every moment, characterized by the values of a variable number of physical quantities. These values are the information that the acquisition system must provide. Assigning a value to a physical quantity is a measurement. The analysis for these fundamental quantities, the complex set of interactions that constitute a process is obtained by a succession of instrumental operations, each one with its own function: the acquisition chain is formed by the ordered set and coordinated various devices providing these functions (see Figure 9.1).

In its basic structure, an acquisition chain should be able, by means of appropriate devices to do the following functions:

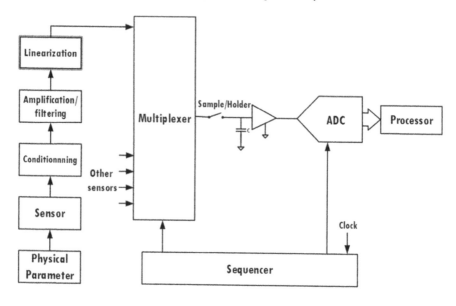

Figure 9.1 Structure of classical digital acquisition chain

- Extracting information for each physical quantity we have to know and translating it into an electrical signal by means of sensors and conditioners.
- Analog signal processing in particular for avoiding degradation by noise: amplification and filtering.
- Linearization of the output signal of the sensor in order to have a linear relationship between the output signal and the measured one, sometimes the linearization device is not in this place if digital linearization is used [21].
- Converting the digital signal through the sample/holder and the ADC.

Once digitized, the data is processed and often, the same material develops a number of orders and data to be transcribed into an analog value in order to act on the outside world.

Linearization is a problem that is often encountered in the measurement systems of physical signals; the reason is that we are facing the effects of nonlinearity in the static characteristic of the sensors. In this sense, we try to have a linear characteristic between the output signal of the sensor and the physical quantity to be measured. For this purpose, a linearization element must be inserted [21, 22].

There are a number of methods of linearization for correcting linearity default of a sensor or its conditioner, when in their area of use; deviations from linearity prohibit considering the sensitivity as a constant for measurement accuracy. These methods can be classified into two classes,

- On one hand, those who work on source of the signal to linearize it from its origin.
- On the other hand, those involved downstream of the source to correct the nonlinearity of the signal which should be done by an analog or digital processing.

The approach presented in this chapter belongs to second class of the linearization methods

9.3 Fuzzy modelling

9.3.1 Fuzzy inference system

Since the introduction of fuzzy sets theory by Zadeh in 1965, it has impressed a wide variety of disciplines and becomes one of the most powerful techniques to modelize the behavior of complex nonlinear systems.

The concept introduced [23], in which fuzzy numbers are assigned to variables to represent uncertainties. A fuzzy value represents the relationship between a vague or uncertain quantity x and a membership function μ, which ranges between 0 and 1 [23]. The inference rule is an "if–then" rule which has the general form "If x is A then y is C", where A and C are linguistic values defined by fuzzy sets in the universes of discourse X and Y, respectively. The if-part is called the antecedent and the then-part is called the consequent of a rule.

The relation $\mu_a(x)$ given by Eq. (9.1) is a fuzzy membership function, which defines the grade of membership of x in A [23]. The most used forms of membership functions are: triangular, trapezoidal, bell shape, Gaussian or sigmoidal functions.

$$A = \{x, \mu_a(x)/x \in X\} \tag{9.1}$$

A fuzzy inference system also known as fuzzy rule-based system or fuzzy model is represented in Figure 9.2 [24, 25]. It is composed of four conventional blocks:

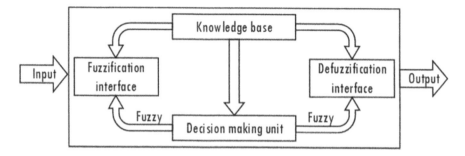

Figure 9.2 Fuzzy inference system

1. Knowledge base unit containing a number of fuzzy if-then rules and defines the membership functions of the fuzzy sets used in the fuzzy rules,
2. Decision-making unit which performs the inference operations on the rules,
3. Fuzzification interface which transforms the crisp inputs into degrees of match with linguistic values,
4. Defuzzification interface which transforms the fuzzy results of the inference into a crisp output.

9.3.2   Takagi-Sugeno fuzzy model

There are two types of fuzzy inference systems: Mamdani models [26] and Sugeno models [25]. The output membership functions of Mamdani models are fuzzy sets, which can incorporate linguistic information into the model, whereas the output membership functions of models are either constant or linear functions of the input variables. Takagi–Sugeno (TS) are more suitable for adaptive modeling in combination with ANNs. This latter technique is going to be used in our design That is why it is suitable to give more details about TS fuzzy modelling. TS rules use functions of input variables as the rule consequent. For fuzzy modelling, a TS rule for three inputs (x_1, x_2 and x_3) and one output (y) is given by Eq. (9.2)

If x_1 is A_1 and x_2 is A_2 and x_3 is A_3 then $y = f(x_1, x_2, x_3)$ (9.2)

Where A_1 , A_2 and A_3 are fuzzy sets in the antecedent, while $y = f(x_1, x_2, x_3)$ is a crisp function in the consequent. Usually the function f is a polynomial in the input variables x and y, but it can be any function as long as it can appropriately describe the output of the model within the fuzzy region specified by the antecedent of the rule. When $f(x, y)$ is a first-order polynomial, the resulting fuzzy inference system is called a first-order Sugeno fuzzy model (see Figure 9.3), which was originally proposed in [25] and [26]. When f is a constant, a zero-order Sugeno fuzzy model is obtained.

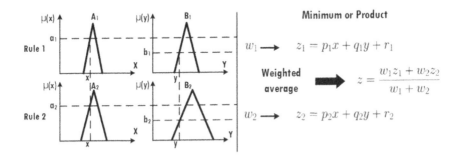

Figure 9.3 First order Sugeno fuzzy model

9.3.3   ANFIS architecture

Since an ANFIS is going to be used, so it's necessary to give a detailed description of the architecture. Figure 9.4 shows the ANFIS architecture where the circular nodes represent fixed operations whereas square nodes represent functions with parameters to be learn. As it seen from Fig. 9.3, ANFIS has five layers and function of these layers can be explained as below [27].

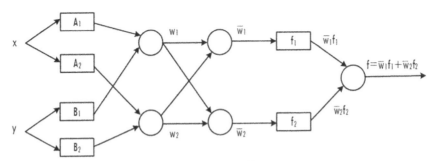

Figure 9.4 ANFIS architecture

- *Layer* 1: In this layer where the fuzzification process takes place, every node is adaptive. Outputs of this layer form the membership values of the premise part. The output of each node is given by Eq. (9.3):

$$O_{1,i} = \mu_{A_i}(x) \quad \text{for } i = 1, 2$$
$$O_{1,i} = \mu_{B_{i-2}}(x) \quad \text{for } i = 3, 4 \tag{9.3}$$

So, the $O_{1,i}(x)$ is essentially the membership grade for x and y.
The membership functions could be triangular, trapezoidal or other type but for illustration purpose we will use the bell shaped function given by Eq. (9.4):

$$\mu_A(x) = \cfrac{1}{1 + \left| \cfrac{x - c_i}{a_i} \right|^{2b_i}} \tag{9.4}$$

Where p_i, q_i, r_i are the parameters to be learned. These are the premise parameters.

- *Layer* 2: The nodes in this layer are fixed. Each node output represents a weight of the rule. Every node in this layer is fixed. This is where the t-norm is used to 'AND' membership grades, for example the product showed by Eq. (9.5):

$$O_{2,i} = w_i = \mu_{A_i}(x)\mu_{B_i}(y) \text{ for } i = 1, 2 \tag{9.5}$$

- *Layer* 3: The nodes are also fixed in this layer. The ration of the ith rule's firing weight to the sum off all rules weights is computed for the corresponding node contains fixed nodes which calculates the ratio of the firing strengths of the rules (Eq. (9.6)):

$$O_{3,i} = \overline{w}_i = \frac{w_i}{w_1 + w_2} \tag{9.6}$$

- *Layer* 4: The nodes in this layer operate as a function block, whose parameters are adaptive and variables are the input values. The output of this layer forms TSK outputs. The nodes in this layer are adaptive and perform the consequent of the rules (Eq. (9.7)):

$$O_{4,i} = \overline{w}_i f_i = \overline{w}_i(p_i x + q_i y + r_i) \tag{9.7}$$

The parameters in this layer (p_i, q_i, r) are to be determined and are referred to as the consequent parameters.

- *Layer* 5: It sums all the incoming signals and produces the output. There is a single node here, and it computes the overall output (Eq. (9.8)):

$$O_{5,i} = \sum_i \overline{w}_i f_i = \frac{\sum_i w_i f_i}{\sum_i w_i} \tag{9.8}$$

This is how, typically, the input vector is fed through the network layer by layer. We now consider how the ANFIS learns the premise and consequent parameters for the membership functions and the rules.

There are a number of possible approaches but we will discuss the hybrid learning algorithm proposed by Jang, Sun and Mizutani which uses a combination of backpropagation and Least Squares Estimation (LSE) [11]. It can be shown that for the network described if the premise parameters are fixed the output is linear in the consequent parameters.

The total parameter set is split into two sets, the set of premise parameters (S_1) which contains the parameters for all the chosen inputs membership functions and the set of consequent parameters (S_2) which contains the parameters of the output membership function.

So, ANFIS uses a two pass learning algorithm; the forward pass where S_1 is unmodified and S_2 is computed using a LSE algorithm, and the backward pass where S_2 is unmodified and S_1 is computed using backpropagation algorithm. Therefore, the hybrid learning algorithm uses a combination of the two cited algorithms to adapt the parameters in the adaptive network.

Model validation is an important step in the modelling step. In practical, the ANFIS is validated with the data remained for the test purpose. Eq. (9.9) gives the root mean square error (*RMSE*) which is the mean criterion used to evaluate the model performance [11].

$$RMSE = \sqrt{\frac{1}{N} \sum_{i=1}^{N} (y_i - \hat{y}_i)^2} \tag{9.9}$$

where N is the number of the tested data, y_i is the ith desired output and $\hat{y}_i$ is the ith predicted output from the model.

9.4 Linearization ANFIS modelling setup

In a data acquisition system, a transducer is said to be linear if corresponding values of its input T_{in}, and output T_{out} lie on straight line. As an example in our application, a temperature measurement sensor is used, it corresponds to a NTC thermistor which is thermally sensitive whose prime function is to exhibit a large, predictable and precise change in electrical resistance when subjected to a corresponding change in temperature. The corresponding characteristic is taken from where more than 1800 data values are considered. As shown in Figure 9.5, the ANFIS block has to learn the inverse characteristic of the sensor.

The developed ANFIS has one input corresponding to the equivalent resistor value (in Volts) and one output corresponding to the Temperature (°C) as illustrated by Figure 9.6. Data experimental values are divided into two parts, 90% of them are used for the identification of the ANFIS parameters and the remaining data are used to test the obtained ANFIS architecture. Using ANFIS toolbox of MATLAB, many approaches are tested with making changes on the type and number of input membership functions and the type of the output membership function, table 1 gives the details of the obtained results in the learning phase. The error calculated by Eq. (9.8), represents the root mean square errors (RMSE) between the ANFIS output and the sensor's inverse characteristic output. To validate the ANFIS obtained model, 180 sets of equivalent resistors values are tested with experimental data that are not used for ANFIS architecture identification. An accuracy of less than 5×10^{-4} is obtained for the developed model in the prediction of the temperature value.

As a digital hardware ANFIS architecture is going to be implemented. So, it's important to choose the right membership number and function type which satisfies the not greedy in hardware resources/linearization precision dilemma. An analysis of the results presented in table 1 leads to choose the second approach (colored in grey on Table 9.1); two input triangle membership functions and two linear output membership functions with

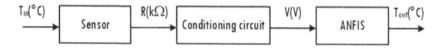

Figure 9.5 ANFIS block in the measurement chain

Figure 9.6 ANFIS linearizer architecture

Table 9.1 Learning phase results

Input Membership		Output Membership	Error	Iterations
Type	*Number*			
	02	Constant	2.7745	150
	02	Linear	1.3065	50
Triangle	03	Constant	1.2863	100
	03	Linear	1.2475	50
	04	Constant	0.9228	50
	04	Linear	0.7386	50
	02	Constant	2.4555	150
	02	Linear	1.3181	50
Trapeze	03	Constant	1.4438	100
	03	Linear	1.3781	100
	04	Constant	1.0290	50
	04	Linear	1.1013	50
	02	Constant	0.8712	500
	02	Linear	0.2669	250
Bell shape	03	Constant	0.4009	100
	03	Linear	0.2137	100
	04	Constant	0.3105	100
	04	Linear	0.1124	50

Table 9.2 Parameters of the input and output membership functions

Input membership functions			Output membership functions		
	a	-1.37000		p	0
Tri_1	b	0.01699	f_1	q	-279
	c	1.44400		r	96.62
	a	0.17400		p	0
Tri_2	b	1.41500	f_2	q	-215.4
	c	2.83000		r	313.7

9.5 FPGA design and implementation

9.5.1 ANFIS circuit digital elements

The final ANFIS architecture obtained after the learning and validations phases can be reduced as illustrated by Figure 9.7. As illustrated by Figure 9.7, the expression of the final output of the ANFIS linearizer is given by (9.10) where the expression of the $Tri_i(x)$ functions ($i = 1$ or 2) are given by (9.11).

$$f(x) = \frac{(q_1 x + r_1)Tri_1(x) + (q_2 x + r_2)Tri_2(x)}{Tri_1(x) + Tri_2(x)} \tag{9.10}$$

$$Tri_i(x) = w_i = \begin{cases} 0 & \text{if } x \leqslant a_i \\ \dfrac{x - a_i}{b_i - a_i} = \dfrac{1}{b_i - a_i}x - \dfrac{a_i}{b_i - a_i} & \text{if } a_i \leqslant x \leqslant b_i \\ \dfrac{c_i - x}{c_i - b_i} = \dfrac{-1}{c_i - b_i}x + \dfrac{c_i}{c_i - b_i} & \text{if } b_i \leqslant x \leqslant c_i \\ 0 & \text{if } c_i \leqslant x \end{cases} \tag{9.11}$$

Hence, the whole ANFIS digital circuit needs:

- Two circuits for the triangle input membership functions ($Tri_1(x)$ and $Tri_2(x)$) which are expressed by (9.11) for $i = 1$ or 2. Where a_i, b_i and c_i are the parameters obtained in the learning phase of the ANFIS (see Table 9.2).
- Two circuits corresponding to the linear output memberships functions ($f_1(x)$ and $f_2(x)$) where the expressions are given by Eq. (9.12) and Eq. (9.13) respectively. q_i and r_i are the parameters obtained in the learning phase of the ANFIS (see Table 9.2).

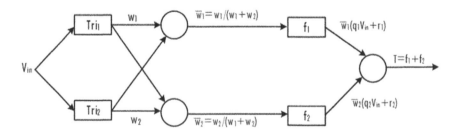

Figure 9.7 Linearization ANFIS architecture

$$f_1(x) = q_1 x + r_1 \qquad (9.12)$$

$$f_2(x) = q_2 x + r_2 \qquad (9.13)$$

9.5.2 FPGA implementation

The functional diagram of the proposed FPGA-based ANFIS is shown in Figure 9.8 where seven different essential blocks can be distinguished. A detailed analysis of expressions given by (9.10), (9.11), (9.12) and (9.13) permits de reduce considerably the consumption of used hardware resources in the implementation phase.

- The ROM block containing the parameters of the input (for $i = 1$ or 2) and output membership functions (a_i, b_i, c_i, p_i, q_i and r_i for $i = 1$ or 2) and the precalculated constants ($1/(b_i - a_i)$, $-a_i/(b_i - a_i)$, $-1/(c_i - b_i)$, $c_i/(c_i - b_i)$) which are computed off line and stored to avoid using additional digital dividers and adders circuits in order to optimize the using of the FPGA hardware resources.
- Two blocks (Tri_1 and Tri_2) corresponding to the input membership functions. Each function needs two multipliers and two adders (see Eq. (9.11)).
- Two blocks (f_1 and f_2) for the output membership functions. Each function needs two multipliers and an adder (see Eq. (9.12) and Eq. (9.13)).
- Two adder circuits for the addition operations needed in Eq. (9.10).
- Two multiplier circuits for the addition operations needed in Eq. (9.10).
- S^{-1} circuit for computing $1/S$ value to S input which needed to compute $(1/(Tri_1(x) + Tri_2(x))$.
- Control unit block: this important circuit controls the proper ordering in the priority of events required by FPGA clock controlling to get the correct functioning of the different blocks at the right moments. This unit operates according to a state machine whose general description diagram is illustrated in Figure 9.9. We can notice that the same adder is used in states 2 and 6, and the same multipliers are used in states 4 and 5.

The adopted representation for the values and signals is 16 bits signed fixed point representation.

The different synthesized modules are described with VHDL. After synthesize, placement and routing phases using Xilinx ISE 10.1i^{TM}

environment, the circuit is implemented in Xilinx Spartan-3A DSP 1800A edition FPGA board.

The full characterization of the circuit should be done by the evaluation of its characteristic. Figure 9.8 shows the block diagram of the experimental test scheme for the synthesized hardware ANFIS based FPGA. A signal generator is used to deliver a signal which is digitized using MAX1132 ADC, the obtained digital signal on Tektronix TLA 5000B logic analyzer permits testing the ANFIS linearizer on the full functioning range.

The values extracted from the data file saved in the logic analyzer are used to trace the graphs of Figure 9.11 and Figure 9.12. Figure 9.11 shows the characteristic of the designed hardware ANFIS for a linear input and the difference between the software linearizer and the designed linearizer is illustrated by Figure 9.12. A maximum error in absolute value of 1.94°C for the hardware ANFIS and 0.98°C for the software ANFIS can be observed.

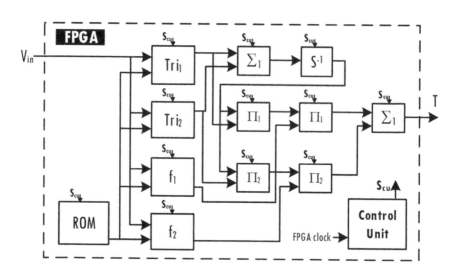

Figure 9.8 Functional diagram of the digital ANFIS

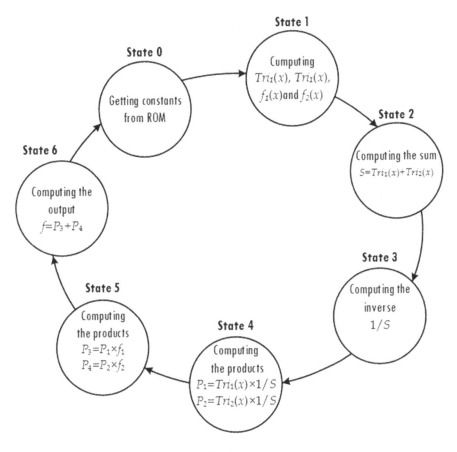

Figure 9.9 State machine diagram of the control unit

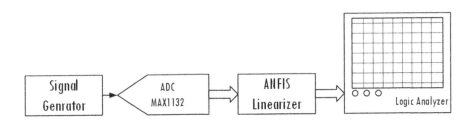

Figure 9.10 Experimental test scheme

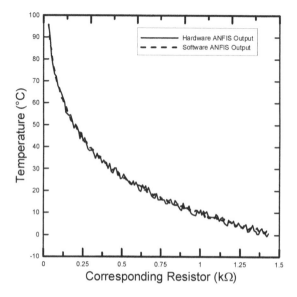

Figure 9.11 ANFIS linearizer response to a linear input

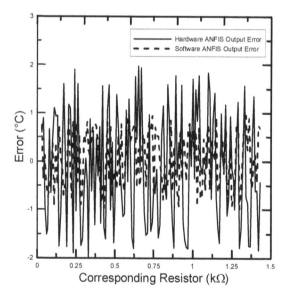

Figure 9.12 Hardware and software ANFIS output errors

For other comparisons, two additional digital circuit based FPGA are used. The first uses look up table technique and the second one uses polynomial interpolation method. The approach proposed in this work consumes less FPGA hardware resources in comparison to the two designed circuits; 40% less than the first one and 55% less than the second one.

9.6 Conclusion

The proposed modular-based design for the ANFIS linearizer is so useful in sensor applications; it will be possible to implement sensors with linearized output digital code. This solution appears to be of lower cost and suitable for VLSI integration, with or without the sensor.

A new optimal neuro-fuzzy architecture is implemented on FPGA, only eight multipliers, six adders, two substractor, S^1 circuit and a small size ROM. The converter has been applied to NTC thermistor to verify its capability. A maximum error in absolute value of 1.94°C is obtained for the hardware designed architecture.

With the same data sets (over than 1800 data values) and for the same precision, the consumption in term of hardware resources using FPGA is reduced to 40% comparing to the technique which uses the look up table (LUT) and 55% with using polynomial interpolation.

The technique presented in this chapter may be further extended to realize others nonlinear ANFIS applications. For further nonlinear sensor applications, changing the ROM values is required to reconfigure the circuit without changing the basic design architecture.

References

[1] R.P. Areny and J.G. Webster, *Sensors and signal conditionning*. New York: Wiley, 2001.

[2] M. Bouhedda and M. Attari, "Synthesis and FPGA-implementation based neural technique of a nonlinear ADC model". *International Scientific Journal of Computing (ISJC)*, Vol. 4, No. 1, pp. 27–33, 2005.

[3] C. Rennenberg, "Analog circuits for thermistor linearization with Chebyshev-optimal linearity error". In *18th European Conference on Circuit Theory and Design (ECCTD'2007)*, Sevilla (Spain), 2007.

[4] H. Erdem, "Implementation of software-based sensor linearization algorithms on low-cost microcontrollers". *ISA Transactions*, Vol. 49, No. 4, pp. 552-558, October 2010.

[5] L. Klopfenstein, "Software linearization techniques for thermocouples, thermistors and

RTDs." *ISA Transactions,* Vol. 33, No. 3, pp. 293–305, September 1994.

[6]   D. Anvekar and B. Sond, "Transducer output signal processing using dual and triple microprocessor systems". *IEEE Transaction on Measurement and Instrumentation,* Vol. 35, No. 2, pp. 182–186, 1986.

[7]   P. Mahana and F. Trofimenkoff, "Transducer output signal processing using an an eight-bit microcomputer". *IEEE Transactions in Instrumentation and Measurement,* Vol. 35, No. 2, pp. 192–186, 1997.

[8]   A. Flamini, D. Marioli and A. Taroni, "Transducer output signal processing using an optimal look-up table in microcontroller-based systems". *Electronics Letters,* Vol. 33, No. 14, pp. 1197–1198, 1997.

[9]   M. Bouhedda, M. Attari and B. Granado, "Hybrid analog FPGA implementation of neural nonlinear ADC for instrument applications". *The Mediterranean Journal of Measurement and Control,* Vol. 3, No. 4, pp. 164–172, 2007.

[10] D. Patranabis, S. Gosh and C. Bakshi, "Linearizing transducer characteristics". *IEEE Transactions on Instrumentation and Measurement,* Vol. 37, No. 1, pp. 66–69, 1988.

[11] J.-S. R. Jang, "ANFIS: Adaptive-network-based inference system". *IEEE Transactions on Systems, Man and Cybernetics,* Vol. 23, No. 3, pp. 665–685, May/June 1993.

[12] M. Attari, M. Boudjema and M. Henniche, "Linearizing a thermistor characteristic in the range of zero to 100°C with two layers artificial neural network". In *IEEE Instrumentation and Measurment Technology Conference (IMTC'1995).* Massachusets (USA): Waltham, 1995.

[13] M. Attari, F. Boudjema and M. Henniche, "An artificial neural network to linearize a G (Tungsten vs. Tingsten 26% Rhenium) thermocouple characteristic in the range of zero to 2000°C". In *IEEE International Symposium on Industrial Electronics (ISIE 1995),* Athens (Greece), July 1995.

[14] M. Meireles, P. Almeida and M. Simoes, "A conprehensive review for industrial applicability of artificial neural networks". *IEEE Transactions on Industrial Electronics,* Vol. 50, No. 3, pp. 585–601, 2003.

[15] P. Daponte and D. Grimaldi, "Artificial neural network in measurement". *Measurement,* Vol. 23, No. 2, pp. 93–115, 1998.

[16] S. Engin, J. Kuvulmaz and V. Ömurlü, "Fuzzy control of an ANFIS model representing a nonlinear liquid-level system". *Neural Computing and Applications,* Vol. 13, No. 3, pp. 202–210, 2004.

[17] P.A.Volnei, *Circuit design with VHDL.* MIT Press, 2004.

[18] Xilinx corporation, "Xilinx data book", 2012 [Online]. Available: www.xilinx.com. [Accessed 18-01-2013].

[19] J. Park and S. Mackay, *Practical Data Acquisition for Instrumentation and Control Systems.* Massachusetts (USA): Elsevier Science, 2003.

[20] G. Asch, *Les capteurs en instrumentation.* Paris (France): Dunod, 1999.

[21] G. Asch, *Acquisition de données: Du capteur à l'ordinateur.* Paris (France): Dunod, 1999.

[22] W. Boyes, *Instrumentation Reference Book*. Massachusetts (USA): Elsevier Science, 2003.

[23] L. Zadeh, "Fuzzy sets", *Information and Control*, Vol. 8, p. 338–353, 1965.

[24] K. Mehran, T. Takagi and K. Sugeno, "Fuzzy modeling for process control", School of Electrical, Electronic and Computer Engineering, Newcastle University, Newcastle, 2008.

[25] T. Takagi and M. Sugeno, "Fuzzy identification of systems and its application to modelling and control". *IEEE Transactions on Systems Man Cybernetics, Part B: Cybernetics*, Vol. 15, pp. 116–132, 1985.

[26] E. Mamdani and S. Assilian, "An experiment in linguistic synthesis with a fuzzy logic controller", *International Journal of Man-Machines Studies*, Vol. 7, pp. 1–13, 1975.

[27] J. Patel and R. Gianchandani, *ANFIS Control for Robotic Manipulators: Adaptive Neuro Fuzzy*. Lap Lambert Academic Publishing, 2011.

[28] U.S. Sensor Corporation, "U.S. sensors", 2013 [Online]. Available: www.ussensor.com/us_sensor_catalog.pdf. [Accessed 08-03-2013].

[29] Z. Huang and J. Hahn, "Fuzzy modeling of signal transduction networks". *Chemical Engineering*, Vol. 64, No. 9, p. 2044–2056, 2009.

Biography

Mounir Bouhedda received the Engineer degree in automatic control engineering from the National Polytechnic School of Algiers (Algeria) in 1996. He received his Master degree in 1999 in robotics and processes control and the Ph.D. degree in 2008 in electronic instrumentation from Houari Boumediene University (USTHB, Algiers, Algeria). He is presently Associate Professor of Electrical Engineering at Médéa University, Algeria from 1999. He teaches courses in the area of electronic instrumentation, automatic control, FGPA implementation and artificial intelligence techniques. He is also a researcher at Advanced Electronic Systems Laboratory (LSEA), Médéa University. He is working on the development and implementation of intelligent instruments and control systems using artificial intelligent techniques using FPGA and microcontrollers solutions.

10

Identification of systems using Volterra model in time and frequency domain

Vitaliy Pavlenko, Sergei Pavlenko and Viktor Speranskyy

ICS, BEITI, Odessa National Polytechnic University, Ukraine

Abstract

The accuracy and noise immunity of the identification method for nonlinear dynamical systems based on the Volterra model in the frequency and time domain is studied. The polyharmonic signals are used as the test ones during identification procedure. The algorithmic and software Matlab toolkit were developed and used for identification process. This toolkit was used to construct the informational model of test system and UHF communication channel. The models are built as a first, second and third order amplitude–frequency and phase–frequency characteristics. The comparison of obtained characteristics with previous works is given. The noise reduction of received characteristics using wavelet denoising were applied and studied.

Keywords: identification, nonlinear dynamic systems, Volterra models, multifrequency characteristics, polyharmonic signals, wavelet denoising, communication channels

10.1   Introduction

One of the most important demands required from communication systems is an accuracy of information transmitted from source to recipient. In natural environment to fulfill this demands we have to eliminate the errors caused by external interference form communication channel (CC) at receiver entrance; internal noise of receiver; signal distortion during transmission. For the last ten years the field related to the methods of signals optimal reception which take into account characteristics of the

V. Haasz and K. Madani (Eds.), Advanced Data Acquisition and Intelligent Data Processing, 233–270.

hardware and CC is intensively developing. The expediency of communication system application depends on how effectively its potential abilities are used. The signals in wireless CC are transferred through a media, i.e. layer of atmosphere, mechanical barriers and so on. The atmospheric constituents cause wavelength dependent absorption and scattering of radiating signal due to environment interactions, emissions and so on (Figure 10.1) [5]. The media between sender and receiver is understood as CC. The technical conditions of CC during operation should be considered for the effective communications. Changes during data transfer can decrease the rate of data transmission in digital CC up to stop of transmission. In analog CC it can be cause distortions and noise of the transmitted signals. Some of the transfer noise effects can be corrected before the received data is converted to information. These effects degrade the adequateness of received data. In marine seaside communications the analog wireless signal can be used as a media for steganographic data [22–23]. The new methods and supporting tools are developed to automate the measurement and consideration of the characteristics of the CC. It helps to build the information and mathematical models of nonlinear dynamical system (NDS) such as the CC [2, 20, 21], i.e. to solve the identification problem.

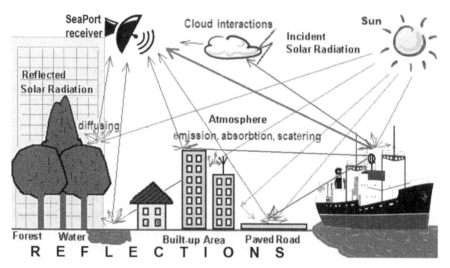

Fig. 10.1 Environment effects in remote sensing systems

Modern continuous CCs are nonlinear stochastic inertial systems. The model in the form of Volterra series used to identify them [2, 3]. The nonlinear and dynamical properties of such system are completely characterized by a sequence of multidimensional weighting functions $w_n(\tau_1,...,\tau_n)$ – Volterra kernels.

Building a model of nonlinear dynamical system in the form of a Volterra series lies in the choice of the test actions form. Also it uses the developed algorithm that allows determining the Volterra kernels $w_n(\tau_1,...,\tau_n)$ and their Fourier-images $W_n(j\omega_1,...,j\omega_n)$ for the measured responses (multidimensional amplitude–frequency characteristics (AFC) and phase-frequency characteristics (PFC)) to simulate the CC in the time or frequency domain, respectively [14–15, 21].

The additional research of new method of nonlinear dynamical systems identification, based on the Volterra model in the frequency domain is proposed. This method lies in n–fold differentiation of responses of the identifiable system by the amplitude of the test polyharmonic signals. The immunity for measurement noise where studied. The wavelet denoising was studied and applied to increase noise immunity of the methods. The developed identification toolkit is used to build information model of the test nonlinear dynamical system in the form of the first, second and third order model [10, 11, 13, 18, 19].

The first aim of this work is reviewing the methods of nonlinear systems identification in frequency and time domain using wavelet filtration of the measurement noises. The next aim is to identify the continuous CC using Volterra model in the frequency domain, i.e. to determinate of its multi-frequency characteristics on the basis of the data of the input-output experiment, using test polyharmonic signals and interpolation method to obtain model coefficients [12, 14–17, 19].

10.2 Volterra models and identification of nonlinear dynamical systems

Generally, input–output type ratio for nonlinear dynamical system can be presented by Volterra series [20]

$$y[x(t)] = w_0(t) + \int_0^\infty w_1(\tau)x(t-\tau)d\tau + \int_0^\infty \int_0^\infty w_2(\tau_1,\tau_2)x(t-\tau_1)x(t-\tau_2)d\tau_1\,d\tau_2 +$$

$$+ \int_0^\infty \int_0^\infty \int_0^\infty w_3(\tau_1,\tau_2,\tau_3)x(t-\tau_1)x(t-\tau_2)x(t-\tau_3)d\tau_1\,d\tau_2\,d\tau_3 + ...$$

$$= w_0(t) + \sum_{n=1}^\infty y_n[x(t)], \tag{10.1}$$

where

$$y_n[x(t)] = \int\limits_0^t \underset{n\ times}{\ldots} \int\limits_0^t w_n(\tau_1,\ldots,\tau_n) \prod_{i=1}^n x(t-\tau_i)d\tau_i \ ,$$

$y_n[x(t)]$ – n–th partial component of response of the system, $x(t)$ and $y(t)$ are input and output signals of system respectively; $w_n(\tau_1,\ldots,\tau_n)$ – weight function or n-order Volterra kernel; $w_0(t)$ – denotes free component of the series (for zero initial conditions $w_0(t)=0$); t – current time.

Commonly, the Volterra series are replaced by a polynomial, with only taking several first terms of series (10.1) into consideration. Nonlinear dynamical system identification in a form of Volterra series consists in n-dimensional weighting functions determination $w_n(\tau_1,\ldots,\tau_n)$ for time domain or it's Fourier transforms $W_n(j\omega_1,\ldots,j\omega_n)$ – n–dimensional transfer functions for frequency domain.

Multidimensional Fourier transform for n-order Volterra kernel (10.1) is written in a form:

$$W_n(j\omega_1,\ldots,j\omega_n) = F_n\langle w_n(\tau_1,\ldots,\tau_n)\rangle = \int\limits_0^\infty \ldots \int\limits_0^\infty w_n(\tau_1,\ldots,\tau_n)\exp\left(-j\sum_{i=1}^n \omega_i\tau_i\right)\prod_{i=1}^n d\tau_i \ ,$$

where $F_n\langle \ \rangle$ – n–dimensional Fourier transform; $j = \sqrt{-1}$. Then the model of nonlinear system based on Volterra model in frequency domain can be represented as:

$$y[x(t)] = \sum_{n=1}^\infty F_n^{-1}\left\langle W_n(j\omega_1,\ldots,j\omega_n)\prod_{i=1}^n X(j\omega_i)\right\rangle_{t_1 = \ldots = t_n = t},$$

where $F_n^{-1}\langle \ \rangle$ – inverse n–dimensional Fourier transform; $X(j\omega_i)$ – Fourier transform of input signal.

Identification of nonlinear system in frequency domain coming to determination of absolute value $|W_n(j\omega_1,\ldots,j\omega_n)|$ and phase $\arg W_n(j\omega_1,\ldots,j\omega_n)$ of multidimensional transfer function at given frequencies – multidimensional AFC and PFC respectively. They are defined by formulas:

$$|W_n(j\omega_1,\ \ldots,j\omega_n)| = \sqrt{[\mathrm{Re}(W_n(j\omega_1,\ldots,j\omega_n))]^2 + [\mathrm{Im}(W_n(j\omega_1,\ldots,j\omega_n))]^2} \ (10.2)$$

$$\arg W_n(j\omega_1,\ldots,j\omega_n) = \operatorname{arctg} \frac{\operatorname{Im}[W_n(j\omega_1,\ldots,j\omega_n)]}{\operatorname{Re}[W_n(j\omega_1,\ldots,j\omega_n)]}, \qquad (10.3)$$

where Re and Im are real and imaginary parts of a complex function of n variables respectively.

So the nonlinear system identification procedure consists in extracting the partial components $y_n[x(t)]$ and determination of multidimensional Volterra kernels or frequency characteristics: AFC and PFC.

10.3 Theoretical foundation of the interpolation method of the system's identification

An interpolation method of identification of the nonlinear dynamical system based on Volterra series is offered [10, 11, 13, 18, 19]. It is used n-fold differentiation of a target signal on parameter-amplitude a of test actions to separate the response of the nonlinear dynamical system on partial components $y_n[x(t)]$.

Statement 1. *Let at input of system test signal of ax(t) kind is given, where x(t) – is arbitrary function and a – is scale coefficient (amplitude of signal), where 0<|a|≤1, then for the selection of a partial component of the n-th order $\hat{y}_n(t)$ from measurement of the response nonlinear system y[ax(t)] in the form of Volterra series, it is necessary to determine n-th partial derivative of the total response amplitude a where a=0*

$$\hat{y}_n(t) = \int\limits_0^t \underset{n \; times}{\ldots} \int\limits_0^t w_n(\tau_1,\ldots,\tau_n) \prod_{i=1}^{n} x(t-\tau_i)d\tau_i = \frac{1}{n!}\frac{\partial^n y[a\,x(t)]}{\partial a^n}\Bigg|_{a=0} \qquad (10.4)$$

Injecting an input signal $ax(t)$ where a is the scale factor (signal amplitude), one has the following response of the nonlinear system:

$$y[a\cdot x(t)] = a\int\limits_0^t w(\tau)\cdot x(t-\tau)d\tau + a^2\int\limits_0^t\int\limits_0^t w_2(\tau_1,\tau_2)\cdot x(t-\tau_1)x(t-\tau_2)d\tau_1 d\tau_2 +$$
$$+ a^n \int\limits_0^t \underset{n\, times}{\ldots} \int\limits_0^t w_n(\tau_1,\ldots,\tau_n)\prod_{r=1}^{n} x(t-\tau_r)d\tau_r +\ldots \qquad (10.5)$$

To distinguish the partial component of the n-th order, differentiate the system response n times with respect to the amplitude:

$$\frac{\partial^n y[a \cdot x(t)]}{\partial a^k} = n! \underbrace{\int_0^t \cdots \int_0^t}_{n\ times} w_n(\tau_1,...,\tau_n) \prod_{r=1}^n x(t-\tau_r)d\tau_r +$$

$$+ (n+1)! \cdot a \underbrace{\int_0^t \cdots \int_0^t}_{n+1\ time} w_{n+1}(\tau_1,...,\tau_{n+1}) \prod_{r=1}^{n+1} x(t-\tau_r)d\tau_r +... \qquad (10.6)$$

Taking the value of the derivative at $a=0$, we finally obtain the expression for the partial component (10.4).

10.3.1 Numerical differentiation for interpolation method of nonlinear systems identification

Partial derivative should be substituted by form of finite difference for calculation. Differentiation of function, which was set in discrete points, could be accomplished by means of numerical computing after preliminary smoothing of measured results.

Various formulas for the numerical differentiation are known, which differ from each other by means of error.

There is an universal way to substitute a derivative of any n order by differential ratio so the error $O(h^p)$ from such substitution for function $y(a)$ would be any beforehand set approximation order p concerning a step $h=\Delta a$ of computational grid by amplitude.

Method of undetermined coefficient for equality is

$$\frac{d^n y(a)}{da^n} = \frac{1}{h^n} \sum_{r=-r_1}^{r_2} c_r y(a+rh) + O(h^p), \qquad (10.7)$$

let's pick up coefficients c_r not depending on h, $r=-r_1, -r_1+1,..., -1,0,1,..., r_2-1, r_2$, so that equality (10.7) was fair. Limits of summation $r_1 \geq 0$ and $r_2 \geq 0$ could be arbitrary, but so that the differential relation $h^{-n} \sum c_r y(a+rh)$ of r_1+r_2 order have to satisfy to inequality $r_1+r_2 \geq n+p-1$.

For definition of c_r it is necessary to solve the following set of equations

$$
\begin{bmatrix}
1 & 1 & \cdots & 1 \\
-r_1 & -r_1+1 & \cdots & r_2 \\
\cdots & \cdots & \cdots & \cdots \\
(-r_1)^{n-1} & (-r_1+1)^{n-1} & \cdots & r_2^{n-1} \\
(-r_1)^{n} & (-r_1+1)^{n} & \cdots & r_2^{n} \\
(-r_1)^{n+1} & (-r_1+1)^{n+1} & \cdots & r_2^{n+1} \\
\cdots & \cdots & \cdots & \cdots \\
(-r_1)^{n+p-1} & (-r_1+1)^{n+p-1} & \cdots & r_2^{n+p-1}
\end{bmatrix}
\cdot
\begin{bmatrix}
c_{-r_1} \\
c_{-r_1+1} \\
\cdots \\
c_{-1} \\
c_0 \\
c_1 \\
\cdots \\
c_{r_2}
\end{bmatrix}
=
\begin{bmatrix}
0 \\
0 \\
\cdots \\
0 \\
n! \\
0 \\
\cdots \\
0
\end{bmatrix} . \quad (10.8)
$$

If $r_1 + r_2 = n + p - 1$, then inscribed in $n + p$ equality forms linear system concerning the same number of c_r unknown. The determiner of this system is Vandermonde's determiner and differs from zero. Thus, there is the only one set of n coefficients, satisfying the system. If $r_1 + r_2 \geq n + p$, then there are many such sets of coefficients c_r.

On the basis of (10.7) the formulas of derivative calculation of the first, second and third orders are received at $a{=}0$ with use of the central differences for equidistant nodes of the computational grid.

Volterra kernel of the first order is determined by formulas as the first derivative at $r_1 = r_2 = 1$, $r_1 = r_2 = 2$ or $r_1 = r_2 = 3$ respectively

$$
\begin{aligned}
y_0' &= \frac{1}{2h}(-y_{-1}+y_1), \\
y_0' &= \frac{1}{12h}(y_{-2}-8y_{-1}+8y_1-y_2), \quad (10.9) \\
y_0' &= \frac{1}{60h}(-y_{-3}+9y_{-2}-45y_{-1}+45y_1-9y_2+y_3).
\end{aligned}
$$

Volterra kernel of the first order is determined by formulas as the first derivative at $r_1 = r_2 = 1$, $r_1 = r_2 = 2$ or $r_1 = r_2 = 3$ respectively

$$
\begin{aligned}
y_0'' &= \frac{1}{h^2}(y_{-1}-2y_0+y_1), \\
y_0'' &= \frac{1}{12h^2}(-y_{-2}+16y_{-1}-30y_0+16y_1-y_2), \quad (10.10) \\
y_0'' &= \frac{1}{180h^2}(2y_{-3}-27y_{-2}+270y_{-1}-490y_0+270y_1-27y_2+2y_3).
\end{aligned}
$$

Volterra kernel of the first order is determined by formulas as the first derivative at $r_1 = r_2 = 2$ or $r_1 = r_2 = 3$ respectively

$$y_0''' = \frac{1}{2h^3}(-y_{-2} + 2y_{-1} - 2y_1 + y_2),$$

$$y_0''' = \frac{1}{8h^3}(y_{-3} - 8y_{-2} + 13y_{-1} - 13y_1 + 8y_2 - y_3).$$

$$(10.11)$$

In the formulas written above, we use the following notations

$$y_0' = y'(0),\ y_0'' = y''(0),\ y_0''' = y'''(0);\ y_r = y(rh),\ r = 0, \pm 1, \pm 2; \pm 3,$$

where we put $y_0 = 0$, since identification nonlinear systems is implemented with zero initial conditions.

The amplitudes of the test signals $a_i^{(n)}$ and the corresponding coefficients $c_i^{(n)}$ for responses are shown in Table 10.1, where n – order of the estimated Volterra kernel; i – number of the experiment ($i = 1, 2, ..., N$), where $N = r_1 + r_2$, i.e. number of interpolation knots (number of experiments).

Table 10.1. Amplitudes and corresponding coefficients of the interpolation method.

n	N	$a_1^{(n)}$	$a_2^{(n)}$	$a_3^{(n)}$	$a_4^{(n)}$	$a_5^{(n)}$	$a_6^{(n)}$	$c_1^{(n)}$	$c_2^{(n)}$	$c_3^{(n)}$	$c_4^{(n)}$	$c_5^{(n)}$	$c_6^{(n)}$
1	2	-1	1	–	–	–	–	-0,5	0,5	–	–	–	–
	4	-1	-0,5	0,5	1	–	–	0,083	-0,667	0,667	-0,083	–	–
	6	-1	-0,67	-0,33	0,33	0,67	1	-0,017	0,15	-0,75	0,75	-0,15	0,0167
2	2	-1	1	–	–	–	–	1	1	–	–	–	–
	4	-1	-0,5	0,5	1	–	–	-0,083	1,333	1,333	-0,083	–	–
	6	-1	-0,67	-0,33	0,33	0,67	1	0,011	-0,15	1,5	1,5	-0,15	0,011
3	4	-1	-0,5	0,5	1	–	–	-0,5	1	-1	0,5	–	–
	6	-1	-0,67	-0,33	0,33	0,67	1	0,125	-1	1,625	-1,625	1	-0,125

The char "–" in Table 10.1 means that there is no value for this amplitude and coefficient.

10.3.2 Identification of dynamical systems as a Volterra model in time domain using poly–impulse signals

Statement 2. *If input test signals x(t) present themselves in irregular serial impulses of different duration and each component consists of not more than of n delta impulses with square of S=aΔτ (Δτ – duration, a – impulse amplitude of squared shape), which function at* $\tau_1,\,\ldots\,,\tau_n$ *time point, then for nonlinear systems with one input and output the estimation of subdiagonal section of Volterra kernels n-th order is:*

$$\hat{w}_n(t-\tau_1,\ldots,t-\tau_n) = \frac{(-1)^n}{n!(\Delta\tau)^n} \sum_{\delta_{\tau_1},\ldots,\delta_{\tau_n}=0}^{1} (-1)^{\sum_{i=1}^{n}\delta_{\tau_i}} \hat{y}_n(t,\delta_{\tau_1},\ldots,\delta_{\tau_n}),\ (10.12)$$

where $\hat{y}_n(t,\delta_{\tau_1},\ldots,\delta_{\tau_n})$ *– is the measured system reaction at time moment t, which was obtained as the result of data processing of experiments based on (10.4), provided that* $\delta_\tau = 1$ *corresponds to an injected impulse at instant* τ*, while* $\delta_\tau = 0$ *means no injected impulse.*

We obtain the estimation of diagonal section of Volterra kernels n-th order

$$\hat{w}_n(t,\ldots,t) = \frac{\hat{y}_n(t)}{(\Delta\tau)^n}, \tag{10.13}$$

where $\hat{y}_n(t)$ *– is a partial component of the n-th order from measurement of the response nonlinear system at single impulse at time point t, which was got as the result of data processing of experiments on bases of (10.4).*

For example, for determination of the lateral section first we test the system by the single impulses acting at instants t_1 and t_2 and find partial components $y_2(t,1,0)$ and $y_2(t,0,1)$. Next, inject the two input pulses:

$$x(t) = S\delta(t\text{-}t_1) + S\delta(t\text{-}t_2) \tag{10.14}$$

where $\delta(t)$ is the Dirac's function.

We separate partial component $y_2(t,1,1)$ from the response of the system to action (10.14) and obtain, according to (10.14), the following expression for the estimation of the Volterra kernel of the second order:

$$\hat{w}_2(t-t_1,t-t_2) = \frac{y_2(t,1,1) - y_2(t,1,0) - y_2(t,0,1)}{2\cdot(\Delta\tau)^2}. \tag{10.15}$$

With fixed values t_1 and t_2, $w_2(t\text{-}t_1,\ t\text{-}t_2)$ represents a function of variable t, precisely, the section of surface $w_2(\tau_1,\tau_2)$ by the plane situated at the angle $45°$ to axes τ_1 and τ_2 and shifted by $\tau_0{=}t_2\text{-}t_1$ along axis τ_1. Changing τ_0 yields various sections $w_2(\tau_1,\tau_2)$ with which one can recover the full surface of the Volterra kernel of the second order.

10.3.3 Identification of dynamical systems as a Volterra model in frequency domain using polyharmonic signals

The test polyharmonic effects for identification in the frequency domain representing by signals of such type:

$$x(t)= \sum_{k=1}^{n} A_k \cos(\omega_k t + \varphi_k),\qquad (10.16)$$

where n – the order of transfer function being estimated; A_k, ω_k and φ_k – amplitude, frequency and a phase of k-th harmonics respectively. In research, it is supposed every amplitude of A_k to be equal, and phases φ_k equal to zero.

For identification in the frequency domain the test polyharmonic signals are used. We prove:

Statement 3. *If used as a test polyharmonic signal*

$$x(t)= A\sum_{k=1}^{n} \cos(\omega_k t)= \frac{A}{2}\sum_{k=1}^{n} \left(e^{j\omega_k t}+e^{-j\omega_k t}\right),\qquad (10.17)$$

then the n–th partial component of the response of test system can be written in form:

$$y_n(t)=\frac{A^n}{2^{n-1}}\sum_{m=0}^{E(n/2)}C_n^m\underbrace{\sum_{k_1=1}^{n}...\sum_{k_n=1}^{n}}_{n}\left|W_n\left(-j\omega_{k_1},...-j\omega_{k_m},j\omega_{k_{m+1}},....j\omega_{k_n}\right)\right|\times$$
$$(10.18)$$
$$\times\cos\left(\left(-\sum_{l=0}^{m}\omega_{k_l}+\sum_{l=m+1}^{n}\omega_{k_l}\right)t+\arg W_n\left(-j\omega_{k_1},...-j\omega_{k_m},j\omega_{k_{m+1}},....j\omega_{k_n}\right)\right),$$

where E – function used to obtain the of integer part of the value. The proving of the statement 3 is written in [8].

The component with frequency $\omega_1+...+\omega_n$ is extracted from the response to test signal (10.16):

$$A^n \, |W_n(j\omega_1, \ldots, j\omega_n)| \cos[(\omega_1 + \ldots + \omega_n)t + \arg W_n(j\omega_1, \ldots, j\omega_n)]. \quad (10.19)$$

Certain limitations should be imposed while choosing of frequency polyharmonic test signals in a process determining multidimensional AFC and PFC. That's why the values of AFC and PFC in this unallowable points of multidimensional frequency space can be calculated using interpolation only. In practical realization of nonlinear dynamical systems identification it is needed to minimize number of such undefined points at the range of multidimensional frequency characteristics determination. This done to provide a minimum of restrictions on choice of frequency of the test signal. It is shown that existed limitation can be weakened. New limitations on choice of frequency are reducing number of undefined points.

After analyzing the (10.19) it is defined: to obtain Volterra kernels for nonlinear dynamical system in frequency domain the limitations on choice of frequencies of test polyharmonic signals have to be restricted. These restrictions provide inequality of combination frequencies in the test signal harmonics. The theorem about choice of test signals frequencies is proven.

10.3.4 The theorem about choosing the test signals frequencies

Theorem. *For the definite filtering of a response of the harmonics with combination frequencies $\omega_1 + \omega_2 + \ldots + \omega_n$ within the n-th partial component it is necessary and sufficient to keep the frequency from being equal to another combination frequencies of type $k_1\omega_1 + \ldots + k_n\omega_n$, where the coefficients $\{k_i | i = 1,2,\ldots,n\}$ must satisfy the conditions:*

- *number K of negative value coefficients $(k_i < 0)$ is in $0 \le K \le E(n/2)$ (where E – function used to obtain the of integer part of the value);*

- $\sum\limits_{i=1}^{n} |k_i| \le n;$

- $\sum\limits_{i=1}^{n} |k_i| \equiv n \,(\text{mod } 2),\ n - \sum\limits_{i=1}^{n} |k_i| = 2l,\, l \in \mathbf{N}.$

It was shown that during determination of multidimensional transfer functions of nonlinear systems it is necessary to consider the imposed constraints on choice of the test polyharmonic signal frequencies.

This provides inequality of combination frequencies in output signal harmonics [9]: $\omega_1 \neq 0$, $\omega_2 \neq 0$ and $\omega_1 \neq \omega_2$ for the second order identification procedure, and $\omega_1 \neq 0$, $\omega_2 \neq 0$, $\omega_3 \neq 0$, $\omega_1 \neq \omega_2$, $\omega_1 \neq \omega_3$, $\omega_2 \neq \omega_3$, $2\omega_1 \neq \omega_2 + \omega_3$,

$2\omega_2 \neq \omega_1 + \omega_3$, $2\omega_3 \neq \omega_1 + \omega_2$, $2\omega_1 \neq \omega_2 - \omega_3$, $2\omega_2 \neq \omega_1 - \omega_3$, $2\omega_3 \neq \omega_1 - \omega_2$, $2\omega_1 \neq -\omega_2 + \omega_3$, $2\omega_2 \neq -\omega_1 + \omega_3$ and $2\omega_3 \neq -\omega_1 + \omega_2$ for the third order identification procedure.

10.4 Simulation of the identification method

The described methods were simulated and studied using computer test identification with use of nonlinear test system.

10.4.1 The techniques of test system identification

A nonlinear test system (Figure 10.2) described by Riccati equation was used for modelling using methods presented earlier:

$$\frac{dy(t)}{dt} + \alpha y(t) + \beta y^2(t) = u(t). \tag{10.20}$$

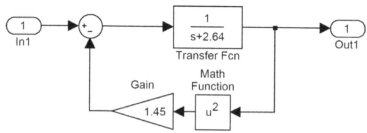

Fig. 10.2 Simulink–model of the test system

Structure chart of nonparametric model of the system is illustrated by means of three Volterra series members in Figure 10.3.

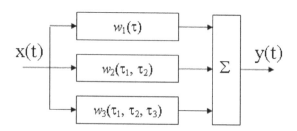

Fig. 10.3 Structure chart of nonparametric model of the system

It is possible to present the Volterra model of chosen test system in the form of kernels of three orders $w_1(\tau), w_2(\tau_1, \tau_2), w_3(\tau_1, \tau_2, \tau_3)$:

$$w_1(\tau) = e^{-\alpha\tau}; \quad w_2(\tau_1, \tau_2) = \frac{\beta}{\alpha}(e^{-\alpha(\tau_1 + \tau_2)} - e^{-\alpha\tau_2}), \quad \tau_1 \le \tau_2;$$

$$w_3(\tau_1, \tau_2, \tau_3) = \frac{1}{3}\left(\frac{\beta}{\alpha}\right)^2 (e^{\alpha(\tau_1 - \tau_2 - \tau_3)} + 3e^{-\alpha(\tau_1 + \tau_1 + \tau_3)} -$$

$$-4e^{-\alpha(\tau_2 + \tau_3)} - 2e^{-\alpha(\tau_1 + \tau_3)} + 2e^{-\alpha\tau_3}), \quad \tau_1 \le \tau_2 \le \tau_3. \tag{10.21}$$

Such diagonal Volterra kernel sections are received having $\tau_1 = \tau_2 = \tau_3 = t$

$$w_2(t,t) = \frac{\beta}{\alpha}(e^{-2\alpha t} - e^{-\alpha t}), \quad w_3(t,t,t) = \left(\frac{\beta}{\alpha}\right)^2 \cdot (e^{-3\alpha t} - 2e^{-2\alpha t} + e^{-\alpha t}).$$

$$\tag{10.22}$$

These analytical expressions are used as etalon characteristics in simulation processes in time domain.

In frequency domain analytical expressions of AFC and PFC for the first, second and third order model were received:

$$|W_1(j\omega)| = \frac{1}{\sqrt{\alpha^2 + \omega^2}}, \quad \arg W_1(j\omega) = -\arctg\frac{\omega}{\alpha};$$

$$|W_2(j\omega_1, j\omega_2)| = \frac{\beta}{\sqrt{(\alpha^2 + \omega_1^2)(\alpha^2 + \omega_2^2)[\alpha^2 + (\omega_1 + \omega_2)^2]}},$$

$$\arg W_2(j\omega_1, j\omega_2) = -\arctg\frac{(2\alpha^2 - \omega_1\omega_2)(\omega_1 + \omega_2)}{\alpha(\alpha^2 - \omega_1\omega_2) - \alpha(\omega_1 + \omega_2)^2} \tag{10.23}$$

$$|W_3(j\omega_1, j\omega_2, j\omega_3)| = \sqrt{[\operatorname{Re}(W_3(j\omega_1, j\omega_2, j\omega_3))]^2 + [\operatorname{Im}(W_3(j\omega_1, j\omega_2, j\omega_3))]^2} =$$

$$= \frac{2\beta^2}{3} \frac{1}{\sqrt{[\alpha^2 + (\omega_1 + \omega_2 + \omega_3)^2][\alpha^2 + \omega_1^2)(\alpha^2 + \omega_2^2)(\alpha^2 + \omega_3^2)}} \times$$

$$\times \frac{\sqrt{\begin{array}{l}[3\alpha^2 - (\omega_1 + \omega_3)(\omega_1 + \omega_2) - (\omega_1 + \omega_2)](\omega_2 + \omega_3) - \\ (\omega_1 + \omega_3)(\omega_2 + \omega_3)]^2 + 16\alpha^2(\omega_1 + \omega_2 + \omega_3)^2\end{array}}}{\sqrt{[\alpha^2 + (\omega_1 + \omega_2)^2][\alpha^2 + (\omega_1 + \omega_3)^2][\alpha^2 + (\omega_2 + \omega_3)^2]}}$$

$$\arg W_3(j\omega_1, j\omega_2, j\omega_3) = \operatorname{arctg} \frac{\operatorname{Im} W_3(j\omega_1, j\omega_2, j\omega_3)}{\operatorname{Re} W_3(j\omega_1, j\omega_2, j\omega_3)} = -\operatorname{arctg} \frac{DA - CB}{AB + CD},$$

where

$$A = 3\alpha^2 - 3\omega_1\omega_2 - 3\omega_2\omega_3 - 3\omega_1\omega_3 - \omega_1^2 - \omega_2^2 - \omega_3^2;$$

$$B = uw - vx; \quad C = 4\alpha(\omega_1 + \omega_2 + \omega_3); \quad D = vw + ux;$$

$$u = \alpha^3 - \alpha\omega_1\omega_3 - \alpha\omega_2\omega_3 - \alpha(\omega_1 + \omega_2 + \omega_3);$$

$$v = (\omega_1 + \omega_2 + \omega_3)(2\alpha^2 - \omega_1\omega_3 - \omega_2\omega_3);$$

$$w = (\alpha^2 - \omega_1\omega_2 - \omega_2\omega_3)(\alpha^2 - \omega_1\omega_2 - \omega_1\omega_3) - \alpha^2(\omega_1 + \omega_2 + \omega_3)^2;$$

$$x = \alpha(\omega_1 + \omega_2 + \omega_3)(2\alpha^2 - 2\omega_1\omega_2 - \omega_1\omega_3 - \omega_2\omega_3).$$

10.4.2   Error estimation of the identification methods

The square criterion is used for error estimation of experimental determination of Volterra kernels sections. To estimate error of experimental determination of Volterra kernels sections the root mean square error (RMSE) is used.

$$\varepsilon = \sqrt{\frac{1}{Q} \sum_{q=1}^{Q} (w_q - \hat{w}_q)^2}, \tag{10.24}$$

where Q – is number of samples at time slice of supervision, w_q – reference values of Volterra kernels, $\hat{w}_q$ – estimation value of kernels received as a result of experimental data (system responses) processing in discrete t_q time points; and also normalized root mean square percentage error (NRMSE)

$$\varepsilon_p = \sqrt{\sum_{q=1}^{Q}\left(w_q - \hat{w}_q\right)^2 \cdot \left(\sum_{q=1}^{Q} w_q^2\right)^{-1}} \cdot 100\ \% \qquad (10.25)$$

10.4.3 Analysis of the test system identification accuracy and noise immunity for interpolation method in time domain

The errors arising during application of determined identification methods are investigated, the comparative analysis of their efficiency on the accuracy and noise stability is carried out. The choice of amplitude of impulses sequence is possible to receive optimum estimates on the accuracy of any Volterra section kernels. The procedures of noise reduction based on wavelet transformations are applied to increase the computing stability of identification algorithms.

Procedures of noise reduction (smoothing) are applied to increase the noise stability of determined identification method to received estimates of the multidimensional Volterra kernels, based on wavelet transformation [4].

Noise reduction is usually reached by removal of high-frequency components from a the signal range representing an additive mix of information component – received as a result of Volterra kernel section processing responses and the noise caused by an error of measuring equipment. In relation to wavelet decomposition it can be realized directly by removal of detailing coefficients of high-frequency levels. It is possible to reduce level of noise by setting wavelet decomposition level and cutting off detailing coefficients respectively.

For smoothing of identification results *wden* utility from a package of the Wavelet Toolbox expansion of MATLAB system [6] with maternal wavelet coiflet – *coif4* was used at the following values of parameters: parameter of the calculation rule for threshold value for restriction of TPTR = 'minimaxi' coefficients (using *minimax* estimation); SORH ='s' (flexible) is for soft or hard thresholding; SCAL = 'one' defines multiplicative threshold without rescaling); depth of data decomposition – 3.

In researches the model of a received noisy estimation of Volterra kernel section is accepted by the additive: $\hat{w}_n(t,\ldots,t) + \xi(t)$ with an even pitch on argument of t, where $\hat{w}_n(t,\ldots,t)$ – is a useful information component, $\xi(t)$ – a hindrance (white Gaussian noise with D dispersion and average zero value).

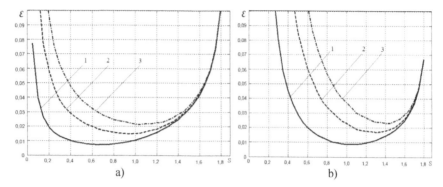

Fig. 10.4 The RMSE dependences of identification using interpolation method for Volterra model of the second (a) and the third orders (b) by the area S of test pulses for $r_1=r_2=1$ and $r_1=r_2=2$ with measurements errors: 1% (1), 3% (2), 5% (3)

There were received dependences (Figure 10.4) of RMSE identification by the area S of test pulse using interpolation method for obtaining the diagonal Volterra kernels sections of the second (Figure 10.4a) and third (Figure 10.4b) orders with measurement error of responses $\sigma = 1$, $\sigma = 3$ and $\sigma = 5\%$ without smoothing procedure of received Volterra kernels sections.

10.4.4 Comparison of the identification methods in time domain

The errors arising during determined identification methods using are investigated. The comparative analysis of its efficiency for accuracy and noise stability is performed. It is possible to choose the amplitudes of impulses sequence to receive optimal estimates for the accuracy of any Volterra kernels sections. The procedures of noise reduction based on wavelet transformations are applied to increase the computing stability of identification algorithms.

The nonlinear dynamical system (Figure 10.2) characteristics received by using three methods of determined identification are given in Table 10.2 – compensation [7], approximation [9] and interpolation methods of diagonal Volterra kernels sections estimation of second order for test, with errors of response measurements $\sigma=1$, $\sigma=3$ and $\sigma=5\%$ without/with wavelet filtration.

Dependence diagrams of RMSE for Volterra kernels identification of second order by the area S (amplitude) of the test impulses for the conditions of ideal experiment (exact measurements) and taking into account errors of measurement responses (error $\sigma=3\%$) are shown in Figures 10.5 and 10.6 respectively.

Table 10.2. Minimum NRMSE for test system identification of the second order Volterra kernels

Parameter of identification methods, N	Minimum NRMSE $\varepsilon_{p_{min}}$ (%) and optimal amplitude Δa_{opt} (relative unit) with error of response measurement σ (%)								
	without wavelet filtration						with wavelet filtration		
	$\sigma = 1$	Δa_{opt}	$\sigma = 3$	Δa_{opt}	$\sigma = 5$	Δa_{opt}	$\sigma = 1$	$\sigma = 3$	$\sigma = 5$
Compensation method									
3	44.0	20	66.5	60	77.1	70	30.1	43.7	53.7
Approximation method									
2	12.6	30	25.9	50	37.0	71	10.8	15.0	18.3
3	11.9	30	24.5	50	33.5	71	9.08	13.3	16.9
4	15.7	55	40.3	75	63.3	83	11.2	18.1	24.5
5	15.2	55	38.0	75	58.7	83	11.1	17.0	22.7
6	18.7	70	50.4	80	80.5	87	11.9	20.5	29.3
Interpolation method									
2	13.0	34	26.3	45	37.5	53	10.9	15.5	19.2
4	14.7	72	36.5	79	58.1	80	11.2	16.8	23.6
6	19.6	84	54.1	86	88.1	87	11.6	20.8	31.5
8	25.6	86	77.3	90	126.0	91	13.1	25.1	44.0

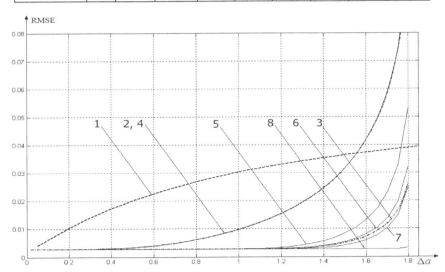

Fig. 10.5 Diagrams of RMSE dependence for diagonal Volterra kernels sections of the second order by the area Δa of test impulses during identification with exact measurements: for a compensation method (1); for interpolation method at $r_1=r_2=1$ (2), at $r_1=r_2=2$ (3); for approximation method at $N=2$, $N=3$ (4), at $N=4$, $N=5$ (5), at $N=6$ (6), at $N=7$ (7), at $N=8$ (8)

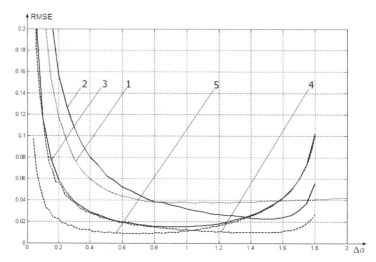

Fig. 10.6 Diagrams of NRMSE dependence for diagonal Volterra kernels section of the second order by the area Δa of test impulses during identification with 3% error measurements: for the compensation method (1); for the approximation method having $N=2$ (2), $N=4$ and $N=5$ (3); for the interpolation method having $r_1=r_2=1$ (4), $r_1=r_2=2$ (5)

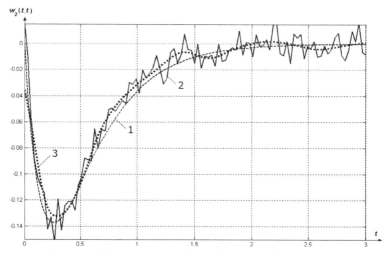

Fig. 10.7 Result of diagonal section identification of the second order using interpolation method ($r_1+r_2=4$) with measurement error 1%: analytically calculated values (1); identified Volterra kernel, NRMSE=7% (2); identified Volterra kernel with wavelet denoising on coif4 wavelet basis with level of decomposition $L=4$, NRMSE=18% (3)

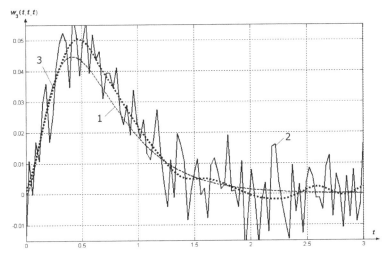

Fig. 10.8 Result of diagonal section identification of the third order using interpolation method ($r_1+r_2=4$) with measurement error 1%: analytically calculated values (1); identified Volterra kernel, NRMSE =7% (2); identified Volterra kernel with wavelet denoising on coif4 wavelet basis with level of decomposition $L=4$, NRMSE=18% (3)

The results of Volterra model building are presented in Figures 10.7 and 10.8. The estimation of diagonal Volterra kernels sections of the second and third orders for the test system with error of responses measurements $\sigma=1\%$ without and with wavelet denoising on the basis of wavelet *coif4* with the decomposition level $L=4$ received using interpolation method of identification.

10.4.5 Analysis of the test system identification accuracy and noise immunity for interpolation method in frequency domain

The main purpose was to identify the multi–frequency properties characterizing nonlinear and dynamical properties of nonlinear test system used in [12]. Volterra model in the form of the 1, 2, 3 order polynomial is used. Thus, test system properties are characterized by transfer functions of $W_1(j\omega)$, $W_2(j\omega_1, j\omega_2)$, $W_3(j\omega_1, j\omega_2, j\omega_3)$ − by Fourier-images of weight functions $w_1(t)$, $w_2(t_1, t_2)$ and $w_3(t_1, t_2, t_3)$.

Structure charts of identification procedure – determinations of the first, second and third order AFC of CC are presented in Figures 10.9–10.11 respectively.

The weighted sum is formed from received signals – responses of each group (Figures 10.9–10.11). As a result the partial components of CC responses $y_1(t)$, $y_2(t)$ and $y_3(t)$ are got. For each partial component of response the Fourier transform (the FFT is used) is calculated, and from received spectrum only an informative harmonics (which amplitudes represent values of required characteristics of the first, second and third orders AFC) are taken.

The first order AFC $|W_1(j\omega)|$ and PFC $\arg W_1(j\omega)$ is received by extracting the harmonics with frequency ω from the spectrum of the CC partial response $y_1(t)$ to the test signal $x(t)=(A/2)\cos\omega t$.

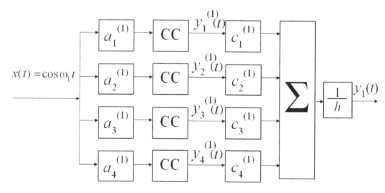

Fig. 10.9 The structure chart of identification using first order Volterra model in frequency domain, number of experiments $N=4$: $a_1=-2h$, $a_2=-h$, $a_3=h$, $a_4=2h$; $c_1=-1/12$, $c_2=-2/3$, $c_3=2/3$, $c_4=1/12$

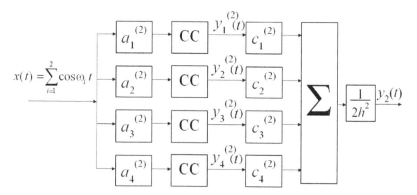

Fig. 10.10 The structure chart of identification using second order Volterra model in frequency domain, number of experiments $N=4$: $a_1=-2h$, $a_2=-h$, $a_3=h$, $a_4=2h$; $c_1=-1/12$, $c_2=4/3$, $c_3=4/3$, $c_4=-1/12$

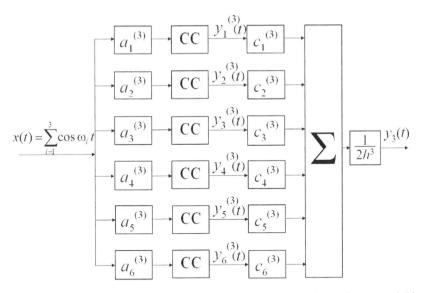

Fig. 10.11 The structure chart of identification using third order Volterra model in frequency domain, number of experiments $N=6$: $a_1=-3h$, $a_2=-2h$, $a_3=-h$, $a_4=h$, $a_5=2h$, $a_6=3h$; $c_1=-1/8$, $c_2=-1$, $c_3=13/8$, $c_4=-13/8$, $c_5=1$, $c_6=1/8$

The second order AFC $|W_2(j\omega, j(\omega+\Omega_1))|$ and PFC $\arg W_2(j\omega, j(\omega+\Omega_1))$ having $\omega_1=\omega$ and $\omega_2=\omega+\Omega_1$ were received by extracting the harmonics with summary frequency $\omega_1+\omega_2$ from the spectrum of the CC partial response $y_2(t)$ to the test signal $x(t)=(A/2)(\cos\omega_1 t+\cos\omega_2 t)$.

The third order AFC $|W_3(j\omega, j(\omega+\Omega_1)), j(\omega+\Omega_2))|$ and PFC $\arg W_3(j\omega, j(\omega+\Omega_1)), j(\omega+\Omega_2))$ having $\omega_1=\omega$, $\omega_2=\omega+\Omega_1$ and $\omega_3=\omega+\Omega_2$ were received by extracting the harmonics with summary frequency $\omega_1+\omega_2+\omega_3$ from the spectrum of the CC partial response $y_2(t)$ to the test signal $x(t)=(A/2)(\cos\omega_1 t+\cos\omega_2 t+\cos\omega_3 t)$.

The results (first, second and third order AFC and PFC) which had been received after procedure of identification are represented in Figures 10.12–10.14.

The surfaces shown in Figures 10.15–10.16 are built from sub–diagonal cross-sections which were received separately. Ω_1 was used as growing parameter of identification with different value for each cross–section in second order characteristics. Fixed value of Ω_2 and growing value of Ω_1 were used as parameters of identification to obtain different value for each cross-section in third order characteristics. The second and third order surfaces for AFC and PFC received after procedure of the test system identification are shown in Figure 10.15 and Figure 10.16 respectively.

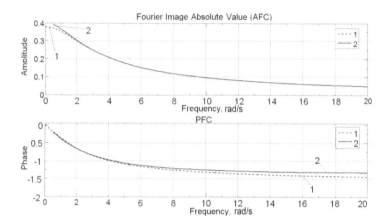

Fig. 10.12 First order AFC and PFC of the test system: analytically calculated values (1), section estimation values with number of experiments for the model $N=4$ (2)

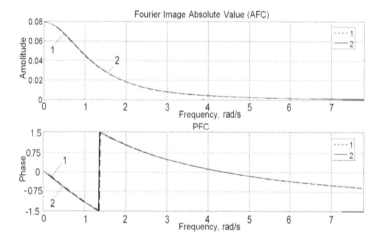

Fig. 10.13 Second order AFC and PFC of the test system: analytically calculated values (1), sub–diagonal cross-section values with number of experiments for the model $N=4$ (2), $\Omega_1=0,01$ rad/s

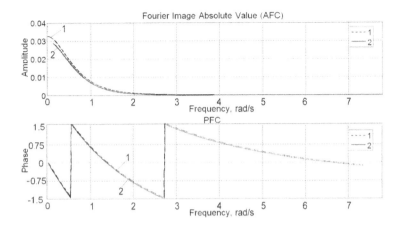

Fig. 10.14 Third order AFC and PFC of the test system: analytically calculated values (1), sub–diagonal cross-section values with number of experiments for the model N=6 (2), Ω_1=0,01 rad/s, Ω_2=0,1 rad/s

Comparison of the numerical values for identification accuracy using interpolation method [10, 11, 18] and approximation one [14–17] for the test system is presented in Table 10.3.

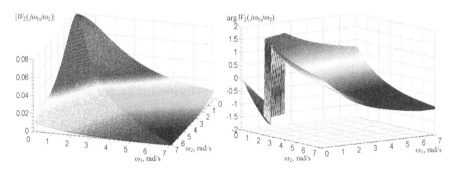

Fig. 10.15 Surface of the test system AFC (left) and PFC (right) built of the second order sub–diagonal cross-sections received for N=4

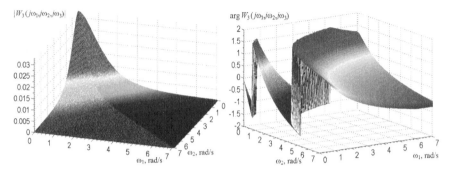

Fig. 10.16 Surface of the test system AFC (left) and PFC (right)built of the third order sub–diagonal cross-sections received for $N=6$, $\omega_3=0,1$ rad/s

Table 10.3. Numerical values comparison for identification accuracy using approximation and interpolation methods

n	N	AFC relative error, %		PFC relative error, %	
		approximation	interpolation	approximation	interpolation
1	2	3.6429	2.1359	3.3451	2.5420
	4	1.1086	0.3468	3.1531	2.0618
	6	0.8679	0.2957	3.1032	1.9311
2	2	26.0092	30.2842	30.2842	76.8221
	4	3.4447	2.0452	2.0452	3.7603
	6	7.3030	89.2099	4.6408	5.9438
3	4	72.4950	10.981	10.9810	1.628
	6	74.4204	10.7642	10.7642	1.5522

10.4.6 Analysis of the test system noise immunity for interpolation method in frequency domain

Experimental researches of the noise immunity of the identification method were made. The main purpose was the studying of the noise impact (noise means the inexactness of the measurements) to the characteristics of the test system model using interpolation method in frequency domain.

The first step was the measurement of the level of useful signal (harmonic cosine test signal shown in Figure 10.17a) after test system (Out2 in Figure 10.18). The amplitude of this signal was defined as the 100% of the signal power.

After that procedure the Random Noise signal (with the form shown in Figure 10.17b) where added to the test system output signal. This was made to simulate inexactness of the measurements in the model. The sum of these two signals for the linear test model signal is shown in Figure 10.19.

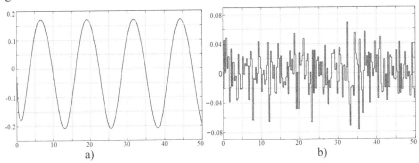

Fig. 10.17 a) Test signal and b) random noise with 50% amplitude of test signal

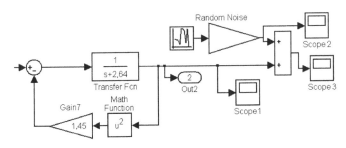

Fig. 10.18 The Simulink model of the test system with noise generator and oscilloscopes

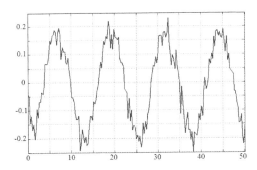

Fig. 10.19 The "noised" signal of the test system, the level of noise is 50% of source signal

The simulations with the test model were made. Different noise levels were defined for different order of the model. The adaptive wavelet denoising was used to reduce the noise impact on final characteristics of the test system. The Daubechie wavelets of the 2 and 3 level were chosen (Figure 10.20) and used for the AFC and PFC denoising respectively [2, 5, 7].

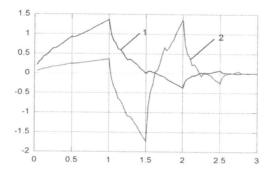

Fig. 10.20 The Daubechies second level scaling (1) and wavelet (2) functions

The first order (linear) model was tested with the level of noise 50% and 10% and showed excellent level of noise immunity. The noised (Figure 10.21a) and denoised (filtered) (Figure 10.21b) characteristics (AFC and PFC) with level of noise 50% are presented.

The second order (nonlinear) model was tested with the level of noise 10% and 1% and showed good level of noise immunity. The noised (Figure 10.22a) and de-noised (filtered) (Figure 10.22b) characteristics (AFC and PFC) with level of noise 10% are presented.

The third order (nonlinear) model was tested with the level of noise 10% and 1% and showed good level of noise immunity. The noised (Figure 10.23a) and de-noised (filtered) (Figure 10.23b) characteristics (AFC and PFC) with level of noise 1% are presented

The numerical values of the standard deviation (SD) of the identification accuracy before/after wavelet denoising procedure are shown in Table 10.4.

The diagrams, showing the improvement of standard deviation for identification accuracy using the adoptive wavelet denoising of the received characteristics (AFC and PFC) are shown in Figures 10.24 and 10.25 respectively.

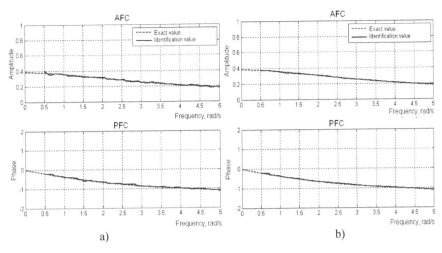

Fig. 10.21 Noised (a) and denoised (b) characteristics (AFC – top, PFC – bottom) of the first order for the test system model with noise level 50%

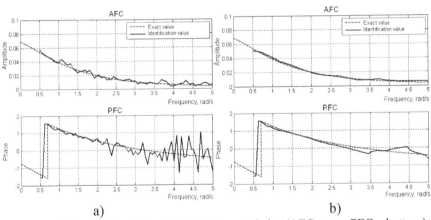

Fig. 10.22 Noised (a) and denoised (b) characteristics (AFC – top, PFC – bottom) of the second order for the test system model with noise level 10%.

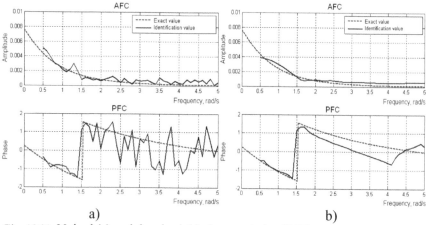

Fig. 10.23 Noised (a) and denoised (b) characteristics (AFC – top, PFC – bottom)
of the third order for the test system model with level of noise 1%

Table 10.4. Standard deviation for interpolation method with noise impact

k	N	Noise level = 10%		Noise level = 1%		Improvement	
		SD for AFC	SD for PFC	SD for AFC	SD for PFC	for AFC, times	for PFC, times
		(without / with denoising)					
1	2	0.000097 / 0.000063	0.09031 / 0.07541	–	–	1,540	1,198
	4	0.000271 / 0.000181	0.07804 / 0.06433	–	–	1,497	1,213
	6	0.000312 / 0.000223	0.12913 / 0.09889	–	–	1,399	1,306
2	2	0.000920 / 0.000670	0.52063 / 0.51465	–	–	1,373	1,012
	4	0.001972 / 0.001663	0.28004 / 0.06877	–	–	1,186	4,072
	6	0.004165 / 0.003908	0.39260 / 0.19237	–	–	1,066	2,041
3	4	–	–	0.000288 / 0.000288	0.89857 / 0.61251	1,003	1,467
	6	–	–	0.000461 / 0.000352	0.84868 / 0.59319	1,310	1,431

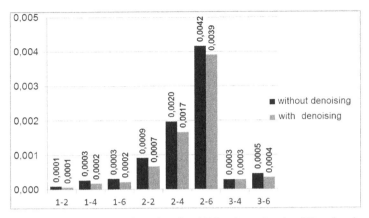

Fig. 10.24 Standard deviation changing for AFC using adoptive Wavelet-denoising

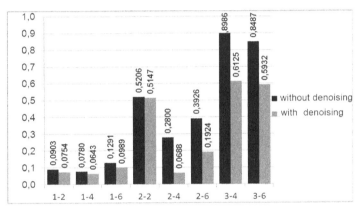

Fig. 10.25 Standard deviation changing for PFC using adoptive Wavelet-denoising

10.5 The algorithm and toolkit of radiofrequency channel identification

Experimental research of the Ultra High Frequency range CC were performed. The main purpose was the identification of multi-frequency characteristics that characterize nonlinear and dynamical properties of the CC. Volterra model in the form of the second order polynomial is used. Thus physical CC properties are characterized by transfer functions of $W_1(j2\pi f)$, $W_2(j2\pi f_1, j2\pi f_2)$ and $W_3(j2\pi f_1, j2\pi f_2, j2\pi f_3)$ – by the Fourier-images of weighting functions $w_1(t)$, $w_2(t_1, t_2)$ and $w_3(t_1, t_2, t_3)$.

Implementation of identification method on the IBM PC computer basis has been carried out using the developed software in Matlab software. The software allows automating the process of the test signals forming with the given parameters (amplitudes and frequencies). Also this software allows transmitting and receiving signals through an output and input section of PC soundcard, to produce segmentation of a file with the responses to the fragments, corresponding to the CC responses being researched on test polyharmonic effects with different amplitudes.

In experimental research two identical marine transceivers S.P.RADIO A/S SAILOR RT2048 VHF (the range of operational frequencies is 154,4−163,75 MHz) and IBM PC with Creative Audigy 4 soundcards were used. Sequentially AFC of the first and second orders were defined. The method of identification with number of experiments $N=4$ was applied. Structure charts of identification procedure − determinations of the first, second and third order AFC of CC are presented in Figures 10.9–10.11 respectively. The general scheme of a hardware–software complex of the CC identification, based on the data of input–output type experiment is presented in Figure 10.26.

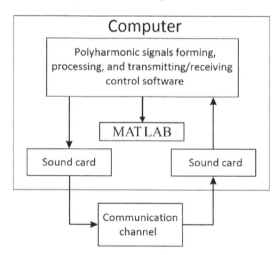

Fig. 10.26 The general scheme of the experiment

The CC received responses $y[a_i x(t)]$ to the test signals $a_i x(t)$, compose a group of the signals, which amount is equal to the used number of experiments N ($N=4$), shown in Figure 10.27. In each following group the signals frequency increases by magnitude of chosen step. A cross-correlation was used to define the beginning of each received response.

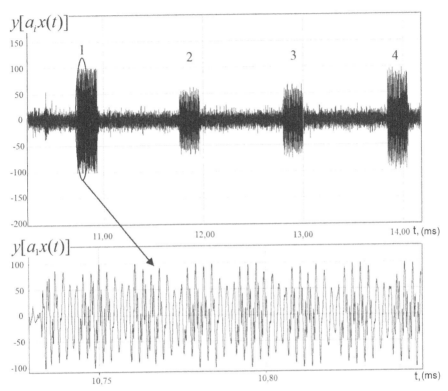

Fig. 10.27 The group of signals received from CC with amplitudes: -1 (1); -1/2 (2); 1/2 (3); 1 (4); $N=4$

Maximum allowed amplitude in the described experiment with use of sound card was $A=0,25$V (defined experimentally). The range of frequencies was defined by the sound card pass band (20...20000 Hz), and frequencies of the test signals has been chosen from this range, taking into account restrictions specified above. Such parameters were chosen for the experiment: start frequency $f_s=125$ Hz; final frequency $f_e=3125$ Hz; a frequency change step $F=125$ Hz; to define AFC of the second order determination, an offset on frequency $F_1=f_2-f_1$ was increasingly growing from 201 to 3401 Hz with step 100 Hz.

The weighed sum is formed from received signals – responses of each group (Figures 10.9–10.11). As a result we get partial component s of response of the CC $y_1(t)$ and $y_2(t)$. For each partial component of the response a Fourier transform (the Fast Fourier Transform is used) is calculated. Only informative harmonics (which amplitudes represents values of required characteristics of the first, second and third order AFC) are taken from received spectrum.

The first order amplitude-frequency characteristic $|W_1(j2\pi f)|$ is received by extracting the harmonics with frequency f from the spectrum of the partial response of the CC $y_1(t)$ to the test signal $x(t)=(A/2)\cos2\pi ft$.

The second order AFC $|W_2(j2\pi f_1, j2\pi f_2)|$, where $f_1=f$ and $f_2=f+F_1$ was received by extracting the harmonics with summary frequency f_1+f_2 from the spectrum of the partial response of the CC $y_2(t)$ to the test signal $x(t)=(A/2)(\cos2\pi f_1 t+\cos2\pi f_2 t)$.

The third order AFC $|W_3(j2\pi f_1, j2\pi f_2, j2\pi f_3)|$, where $f_1=f$, $f_2=f+F_1$ and $f_3=127,5$ Hz were received by extracting the harmonics with summary frequency $f_1+f_2+f_3$ from the spectrum of the partial response of the CC $y_3(t)$ to the test signal $x(t)=(A/2)(\cos2\pi f_1 t+\cos2\pi f_2 t+\cos2\pi f_3 t)$.

The wavelet noise-suppression was used to smooth the output data of the experiment [4]. The results received after digital data processing of the data of experiments (wavelet "Coiflet" denoising) for the first, second and third order AFC are presented in Figures 10.28–10.31.

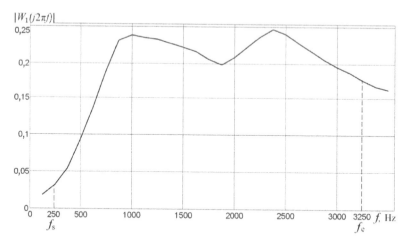

Fig. 10.28 First order AFC after wavelet "Coiflet" second level noise-suppression

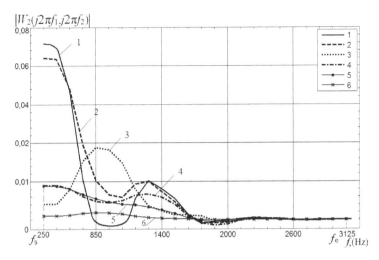

Fig. 10.29 Subdiagonal sections of second order AFCs after wavelet "Coiflet" second level noise-suppression at different frequencies F: 201 (1), 401 (2), 601 (3), 801 (4), 1001 (5), 1401 (6) Hz

The surfaces shown in Figures 10.30–10.31 are built from subdiagonal sections that are received separately. We used F_1 as growing parameter of identification with different value for each section.

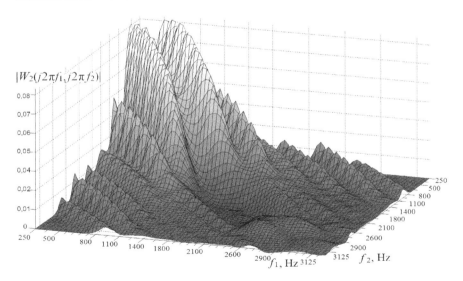

Fig. 10.30 Surface built of the second order AFCs after wavelet "Coiflet" third level noise-suppression

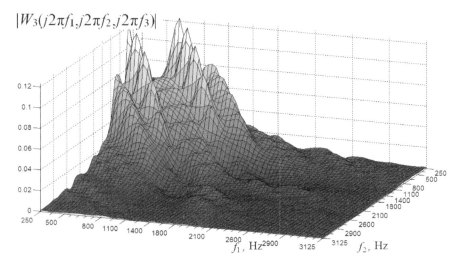

Fig. 10.31 Surface built of AFCs of the third order after wavelet "Coiflet" third level noise-suppression, where f_3= 127,5 Hz

10.6   Conclusion

Methods of the determined identification of nonlinear dynamic systems on the basis of Volterra models in time domain with use irregular sequences of impulses as test signals and in frequency domain with use sequences of polyharmonic signals are studied.

Computational algorithms of Volterra kernels identification for systems with one entrance and exit, and also for systems with many entrances and many exits are shown. It is shown that choosing the sequence impulses parameters – duration, amplitudes and delays between impulses, it is possible to receive optimum estimates on the accuracy of any diagonal and sub-diagonal Volterra kernels sections. In frequency domain it is possible to choose optimal test signals amplitudes and concerning coefficients. And it is necessary to keep the limitations on test signals frequencies.

Comparison of identification results by three methods by means of irregular impulses sequences on test object shows that the highest precision and noise stability the interpolation method of identification consisting in differentiation of responses on parameter amplitude of trial impulses possesses. Least exact of considered methods of the determined identification is the compensation method. The approximating method which is based on drawing up linear combinations of responses of system on test sequences of impulses with different amplitudes, has high rates of efficiency, though concedes on accuracy to an interpolation method of identification.

The methods based on Volterra model using polyharmonic test signals for identification nonlinear dynamical systems are analyzed. To differentiate the responses of system for partial components we use the interpolation method based on linear combination of responses on test signals with different amplitudes.

New values of test signals amplitudes were defined and tested. They are greatly raising the accuracy of identification compared to amplitudes and coefficients written in [1, 12, 14-16]. The accuracy of identification of nonlinear part of the test system growth 2 times and the standard deviation in this case is about 5%.

Pilot studies of the errors arising with application of determined identification methods by means of computer modeling in the MATLAB–SIMULINK environment for the example of system with square nonlinearity in feedback are executed.

Errors of measurements have essential impact on results of the determined identification. Noise reduction procedures using wavelets are applied to increase of computing stability of identification considered algorithms to received estimates of multidimensional Volterra kernels, based on wavelet-transformation. It allows to receive smoothed decisions and reduce identification error in 1,5–2 times.

Communication channel functioning as a media for wireless communication systems is analyzed. Nonlinear effects of the environments have great impact on transfer data received in experiments. The interpolation method of identification using the hardware methodology written in [12, 14] is applied for construction of informational Volterra model as an APC of the first and second order for UHF band radio channel. Received results reveal essential nonlinearity of the CC that leads to distortions of signals in radio broadcasting devices, reduces the important indicators of the TCS: accuracy of signals reproduction, throughput, noise immunity.

The noise immunity is very high for the linear model, high enough for the second order nonlinear model and has moderate noise immunity for the third order model. The wavelet denoising is very effective and makes the gives the possibility to improve the quality of identification of the noisy measurements up to 1,54 and 4,07 times for the AFC and PFC respectively. Final characteristics of the CC have to be used to maintain systems to improve the adequateness of received data.

References

[1] Donoho D.L. and Johnstone I.M. Threshold selection for wavelet shrinkage of noisy data. In *Proc. 16th Annual Conf. of the IEEE Engineering in Medicine and Biology Society*, pp. 24a–25a, IEEE Press, 1994.

[2] Doyle F.J., Pearson R.K., and Ogunnaike B.A. Identification and control using Volterra models. *Springer Technology & Industrial Arts*, 2001.

[3] Giannakis G.B. and Serpedin E.A. Bibliography on nonlinear system identification and its applications in signal processing, communications and biomedical engineering. *Signal Processing, EURASIP,* 81(3), pp. 533–580, 2001.

[4] Goswami J.G. and Chan A.K. *Fundamentals of Wavelets: Theory, Algorithms, and Applications*. John Wiley & Sons, 1999.

[5] Liew S.C. *Principles of remote sensing. Space View of Asia* (second edition). CRISP, 2001. *http://www.crisp.nus.edu.sg/~research/tutorial/ rsmain.htm* (accessed 1 November 2013).

[6] Misiti M., Misiti Y., Oppenheim G., and Poggi J-M. *Wavelets Toolbox Users Guide*. The MathWorks. Wavelet Toolbox, for use with MATLAB, 2000.

[7] Pavlenko V.D. The compensation method for identifying the nonlinear dynamical systems in the form of Volterra kernels. *Trans. of Odessa Polytech. Univ.*, Vol. 2(32), pp. 121–129, 2009.

[8] Pavlenko V.D. and Issa S.I. Limitation in selecting polyharmonic test signals frequences for identifying nonlinear systems. *Trans. of Odessa Polytech. Univ.*, Vol. 1(31), pp. 107–113, 2009.

[9] Pavlenko V., Speranskyy V., Ilyin V., and Lomovoy V. Modified approximation method for identification of nonlinear systems using Volterra models in frequency domain. In *Applied Mathematics in Electrical and Computer Engineering, Proceedings of the AMERICAN–MATH'12 & CSST'12 & CEA'12*, Harvard, Cambridge, USA, January 25–27, 2012. WSEAS Press, pp. 423–428, 2012.

[10] Pavlenko V.D., Pavlenko S.V., and Speranskyy V.O. Interpolation method of nonlinear dynamical systems identification based on Volterra model in frequency domain. In *Proceedings of the 7th IEEE International Conference on Intelligent Data Acquisition and Advanced Computing Systems: Technology and Applications (IDAACS'2013)*, 15–17 September 2013, Berlin, Germany, pp. 173–178, 2013.

[11] Pavlenko V.D. and Speranskyy V.A. Analysis of identification accuracy of nonlinear system based on Volterra model in frequency domain. *American Journal of Modeling and Optimization*, Vol. 1, No. 2, pp. 11–18, 2013. DOI: 10.12691/ajmo-1-2-2.

[12] Pavlenko V.D. and Speranskyy V.O. Communication channel identification in frequency domain based on the Volterra model. Recent advances in computers, communications, applied social science and mathematics. In *Proceedings of the International Conference on Computers, Digital Communications and Computing (ICDCC'11)*, Barcelona, Spain, September 15-17, 2011. WSEAS Press, pp. 218–222, 2011.

[13] Pavlenko V.D. and Speranskyy V.O. Interpolation method modification for nonlinear objects identification using Volterra model in frequency domain. In *Third International Crimean Conference "Microwave & Telecommunication Technology" (CriMiCo'2013)*, Sevastopol, Ukraine, pp. 257–260, 2013.

[14] Pavlenko V.D., Speranskyy V.O., and Lomovoy V.I. Modelling of radio-frequency communication channels using Volterra model. In *Proc. of the 6th IEEE International*

Conference on Intelligent Data Acquisition and Advanced Computing Systems: Technology and Applications (IDAACS'2011), 15–17 September 2011, Prague, Czech Republic, pp. 574–579, 2011.

[15] Pavlenko V.D., Speranskyy V.O., and Lomovoy V.I. The test method for identification of radiofrequency wireless ommunication channels using Volterra model. In *Proc. of the 9th IEEE East–West Design & Test Symposium (EWDTS'2011)*, Sevastopol, Ukraine, September 9–12, 2011. Kharkov: KNURE, pp. 331–334, 2011.

[16] Pavlenko V., Lomovoy V., Speranskyy V., and Ilyin V. Radio frequency test method for wireless communications using Volterra model. In *Proc. of the 11th Conference on Dynamical Systems Theory and Applications (DSTA'2011)*, December 5–8, 2011, Łódź, Poland, pp. 446–452, 2011

[17] Pavlenko V.D. and Speranskyy V.O. Analysis of nonlinear system identification accuracy based on Volterra model in frequency domain. *Electrotechnic and Computer Systems*, Vol. 8(84), pp. 66–71, 2012.

[18] Pavlenko V.D. and Speranskyy V.O. Identification of nonlinear dynamical systems using Volterra model with interpolation method in frequency domain. *Electrotechnic and Computer Systems*, Vol. 5(81), pp. 229–234, 2012.

[19] Pavlenko V.D. and Speranskyy V.O. Simulation of telecommunication channel using Volterra model in frequency domain. In *Proceedings 10th IEEE East–West Design & Test Symposium (EWDTS'2012)*, Kharkov, Ukraine, September 14–17, 2012, pp. 401–404.

[20] Schetzen M. *The Volterra and Wiener Theories of Nonlinear Systems*. New York: Wiley & Sons, 1980.

[21] Westwick D.T. Methods for the identification of multiple-input nonlinear systems. Departments of Electrical Engineering and Biomedical Engineering, McGill University, Montreal, Quebec, Canada, 1995.

[22] Shyshkin A.V., Koshevoy V.M. Steganographic data transmission robust against scaling attacks (resistant-to-scaling-attacks) and eliminating the hampering effect of signal-carrier. *Izv. Vyssh. Uchebn. Zaved., Radioelektron.*, Vol. 50(6), p. 3, 2007.

[23] Shishkin O.V. and Koshevyy V.M. Audio watermarking for automatic identification of radiotelephone transmissions in VHF maritime communication. *Watermarking. Published InTech*, pp. 209–228, 2012. DOI: 10.5772/36851.

Biographies

Vitaliy Pavlenko (PhD, DSc) was awarded an engineering diploma (five-year degree program) in 1970 and a PhD degree in 1987 at the Odessa State Polytechnic University (in the former USSR). He was awarded a DSc degree in 2013 from the same university (Ukraine). Prof. Pavlenko is currently with Odessa National Polytechnic University, Ukraine. His research interests include among others modeling and simulation for industrial applications, non-parametric identification of nonlinear systems, theory of the Volterra series, mathematical methods, models and technologies for complex systems' research.

Sergey Pavlenko was awarded a MEng degree at Odessa National Polytechnic University in 2007. He is working as a junior researcher at the Energy Management Institute and is preparing his PhD thesis now. His research activities are related to modeling and simulation, non-parametric identification of nonlinear systems and theory of the Volterra series.

Viktor Speranskyy received a MEng degree in 2003 from Odessa State Academy of Refrigeration. He is currently working towards a PhD degree in the Computer Systems Institute, Odessa National Polytechnic University. Simultaneously he is currently a senior lecturer of Economic cybernetics and informational technologies department in the Business, Economics and Informational Technologies Institute. His research activities are pointed at modeling and simulation, non-parametric identification of nonlinear systems and theory of the Volterra series.

11

Training cellular automata for hyperspectral image segmentation[*]

B. Priego, D. Souto, F. Bellas and R. J. Duro

Integrated Group for Engineering Research, Universidade da Coruña, Spain

Abstract

This chapter describes a general approach for the segmentation of hyperspectral images that provides a solution to three of the main problems within the area. On one hand, using Cellular Automata (CA) it permits an easy parallelization of the computations required, thus reducing the very high computational cost this type of tasks involve. On the other, and more importantly, it addresses the problem of the lack of adequately labelled hyperspectral images, which are necessary for the evaluation of candidate solutions. This is achieved by allowing the CAs rule sets to be obtained through evolution by just using adequately chosen synthetic images of a much lower dimensionality than those over which they will be run. Finally, this approach uses the whole wealth of spectral information in the images by providing a CA workflow that does not project the spectral images onto lower dimensions in any point of the process. In other words, it always works with the complete hyperspectral image. The effectiveness of the proposed approach has been verified over the segmentation of synthetic and real-world hyperspectral images leading to very competitive results.

Keywords: hyperspectral imaging, evolution, cellular automata

---

[*] This work was partially funded by the Spanish MICINN and European Regional Development Funds through project TIN2011-28753-C02-01.

V. Haasz and K. Madani (Eds.), Advanced Data Acquisition and Intelligent Data Processing, 271–292.

11.1 Introduction

Hyperspectral imaging is nowadays a very relevant field for remote sensing and industrial processing applications where managing data with a very high level of spectral detail provides a clear advantage. In this sense, hyperspectral images provide hundreds of values for each pixel that correspond to the intensity of a large number of bands within the visible and near infrared spectrum, thus facilitating discrimination of different elements within the images. They are typically made up of two spatial dimensions of anywhere from 250 pixels to a few of thousand pixels and a spectral dimension of up to a thousand bands per pixel.

As in the case of RGB imaging, one of the most important operations that can be performed over images is segmentation, that is, the discrimination of areas that correspond to the same label (which depends on the application). In this case, segmentation of hyperspectral images implies the integration of spatial and spectral information using data from the spectrum of the pixel and of the pixels around it in order to decide to which class this pixel belongs. Most approaches that address this problem are basically extensions of traditional RGB image processing techniques, such as region-merging [2] or hierarchical segmentation [16]. Some of these approaches, such as morphological levels [12] or Markov random fields [5, 6], apply fixed window based neighbourhoods, which reduces their flexibility. Others are based on mathematical morphology techniques [7, 11] or watershed based algorithms [15, 16], trying to extend them to this domain. In these cases an ordering problem arises that is typical of high dimensional spaces, which makes the adaptation of these techniques quite complex. Some advanced approaches apply intelligent learning algorithms, such as those presented in [4, 10, 18]. They usually require complex algorithms that lead to computationally intensive processing stages that are very hard to implement in limited computing resources.

The spectral and spatial details contained in hyperspectral images imply the necessity of dealing with very large data sets. This represents a severe processing problem that requires the development of specialized techniques and the use of very high levels of parallelism. Recently, different authors have started to face this problem by means of distributed architectures based on FPGAs [2] or GPUs [9], especially when real time processing is desired. But adapting the previously mentioned algorithms to parallel execution is not easy and very specialized ad-hoc implementations have to be developed in order to benefit from their architecture.

Another two problems in hyperspectral imaging segmentation must be faced in order to be able to achieve good performance levels. The first one comes from the reliability of the labelled sets of images that can be used as ground truth, which is usually quite poor, not to mention the difficulty in creating large sets of these images. Consequently, it is necessary to provide methodologies that allow performing this type of processing without having to resort to large training or labelled sets of hyperspectral images. The second one is related to the necessity of preserving the representational power of these high dimensional images. In this sense, it is important that the algorithms developed work with the complete spectral wealth of the images and do not project them onto lower dimensionality spaces.

To address these problems, we have first developed a segmentation algorithm that is based on a Cellular Automata (CA) [8, 14] in order to create spectrally representative relatively homogeneous regions. The CA rule definition may become highly complex in this type of images, so we decided to apply an evolutionary algorithm for their automatic design. The resulting algorithm has been called ECAS (Evolutionary Cellular Automata based Segmentation) and it has provided preliminary successful results when compared to similar approaches [14]. As CA based techniques are intrinsically distributed, ECAS can be easily executed in parallel thus reducing the computational cost problem. Moreover, it allows performing segmentation and classification processes using only spectrally relevant information, that is, without irreversibly projecting the images onto lower dimensional spaces. Finally, as the evolutionary process requires an evaluation phase that needs some type of ground truth, instead of resorting to the poorly labelled hyperspectral images commented above, we have developed a training technique for the CA that is carried out over synthetic RGB images. It has been shown that CAs trained this way and taking into account all the details of the methodology can then be directly applied to hyperspectral ones.

The rest of the chapter is structured as follows: section 11.2 is devoted to the definition and basic operation of the ECAS algorithm, focusing the attention on the training problem. Sections 11.3 and 11.4 contain two examples of application of the algorithm in two real hyperspectral images used as benchmark cases in the field that show the validity of the proposed approach. Finally, section 11.5 contains the discussion and main conclusions of this chapter.

11.2 Evolutionary cellular Automata based segmentation

11.2.1   CA definition

The approach followed here involves the application of cellular automata in order to segment and classify hyperspectral images. In this case, the cellular automaton consists in a two-dimensional grid of cells, so that every cell is placed over each pixel of the hyperspectral image. Due to this one to one relationship, the terms pixel and cell will be considered interchangeable.

The state of the cell (S_i) is given by an N-band spectrum, taking values in the range [0, 1]. Consequently, the state space is continuous and corresponds to the positive vector space in $\mathbb{R}^N$. The cellular automaton is applied to the whole hyperspectral cube iteratively, gradually modifying the state or spectrum of every pixel of the image. How the state of a cell is modified in each iteration depends on the spectrum in that location, on the spectra of the eight closest neighbouring cells (Moore neighbourhood [7]) (S_{i1}, S_{i2}, ..., S_{i8}) and on the set of transition rules that controls the automaton behaviour. Due to the continuity of the state space, we will speak of state modifications rather than state changes. It is important to note here that the procedure presented here preserves the spectral information, that is, it works with the whole N-band state vector throughout the segmentation process.

In order to apply the transition rules that will somehow modify the spectra that define the state of the cells, the CA needs to collect information about its neighbourhood. Specifically, to compare the state or spectrum of neighbouring cells, it is necessary to establish a distance measure. In this case, the spectral angle (normalized between 0 and 1) has been the selected measure. For a cell i, the normalized spectral angle α_{ij} with respect to cell j is defined as:

$$\alpha_{ij} = \frac{2}{\pi}\cos^{-1}\left(\frac{\sum s_{ij}s_i}{\sqrt{\sum s_{ij}^2}\sqrt{\sum s_i^2}}\right), \quad \text{for } j=1, 2, ..., 8 \qquad (11.1)$$

For each cell of the hyperspectral image, a vector $A_i=(\alpha_{i1}, \alpha_{i2}, ..., \alpha_{i8})$ made up of the eight angles with respect to the eight neighbouring cells, can be defined.

The rule set of the cellular automaton consists of two threshold angles, ε and θ, and a set of M rules, each one of them made up of 9 parameters r_{xj}:

$$CA = (\varepsilon, \theta, r_{11}, \ldots, r_{19}, r_{22}, \ldots, r_{29}, \ldots, r_{M1}, \ldots, r_{M9}) \qquad (11.2)$$

with:

$$\begin{cases} 0 < \varepsilon < 1 \\ \varepsilon < \theta < 1 \\ r_{ij} = \{-1, 0, 1\} \\ j = 1, 2, \ldots, 8 \\ r_{i9} = \{1, 2, \ldots, 8\} \end{cases} \qquad (11.3)$$

Every iteration of the CA, only one of these M rules is applied over each pixel. To decide which one is selected, a very simple procedure involving the comparison of neighboring spectral angles has been developed. Basically, vector A_i for the cell is obtained and discretized (Figure 11.1) into a set of discretized angles (α_{cij}) depending on the threshold angles ε and θ as:

$$\alpha_{C_{ij}} = \begin{cases} -1 & \text{if} \quad \alpha_{ij} < \varepsilon \\ 0 & \text{if} \quad \varepsilon < \alpha_{ij} < \theta \\ 1 & \text{if} \quad \alpha_{ij} > \theta \end{cases} \qquad (11.4)$$

The automaton selects the rule whose first eight r values ($r_{x1}, r_{x2}, \ldots, r_{x8}$) are most similar in terms of Euclidean distance to the discretized A_i and applies it to modify the state of cell i. The ninth parameter of each rule is an output that indicates a cell neighbor that is selected. This parameter is encoded as an integer value between 1 and 8 corresponding to one of the neighbors following the directions NW, N, NE, W, E, SW, S and SE. In other words, when a CA selects a rule j for application and its ninth parameter is 5, this means that the spectrum of cell i will be modified in the direction of that of the cell on its left, becoming more similar to it.

As a first approximation, a spectrum-averaging operator will be considered that implies that the resulting spectrum is the band by band average of the two spectra. In addition, to make rule selection invariant to rotations and reflections, four possible rotations and the vertical and horizontal flips of each one of them are taken into account during the rule selection process.

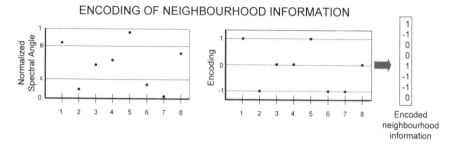

Fig. 11.1. Example of encoding of the neighborhood information

The cellular automaton is iteratively applied to all the cells of the image producing an updated hyperspectral cube every iteration.

11.2.2   Segmentation

Although the key aspect of the ECAS algorithm is in the CA, to test its performance, a segmentation algorithm is required. After N iterations of applying the cellular automaton to all the pixels of the image, the resulting hyperspectral cube is a modified version of the original one, in which regions that should be segmented with the same label have been spectrally homogenized.

Two different procedures have been followed in order to demonstrate the advantages of using the homogenized version of the data cube to segment or classify the hyperspectral image. Firstly, the hyperspectral image has been segmented by means of applying a k-means algorithm to every pixel (N-element vector) fixing the number of clusters considered in the segmentation. This procedure is used in the experiments presented in the application example of section 11.4. In the second procedure, the new data cube is segmented using a border detection algorithm that determines the limits of each area of the image. Then, to produce a provisional classification image, a SVM based pixel-wise classification is applied to the original hyperspectral cube following the work of Tarabalka et al. [16] A multi-class pairwise (one versus one) SVM classification of the hyperspectral image is performed using a Gaussian Radial Basis Function (RBF) kernel. The specific implementation used was the one found in the LIBSVM library 0, using the C and γ parameter values of [16]: $C = 128$ and $\gamma = 0.125$. These parameters will be considered constant for all the experiments presented in section 11.5.

The result of this provisional classification is combined with the resulting segmented image from the CA in order to obtain a final classification image. All the pixels in an area of the segmented image are labelled using the most frequent label in the provisional classification result for the same area. Figure 11.2 displays a diagram of the whole segmentation-classification process followed in this second procedure.

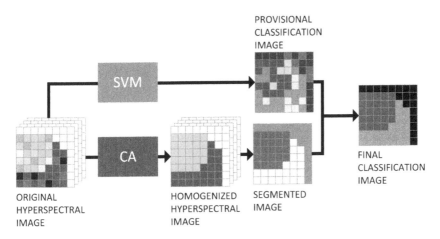

Fig. 11.2. Segmentation and classification process diagram

11.2.3 CA training

The main difficulty of designing cellular automata to perform a particular task lies in selecting the set of transition rules that make up the cellular automata. This selection is a complex problem to solve manually. For this reason, in this work the generation of the transition rules of the cellular automaton is addressed in an automatic manner though the application of an evolutionary algorithm (EA).

There are several reasons for choosing EA as the optimization method in the selection of the transition rules. First, EAs do not make assumptions about the underlying fitness landscape. Second, EAs work on a population of solutions instead of a single point. This makes it less likely for them to be trapped in a local minimum in the search for the optimum set of transition rules.

EAs operate over a population of individuals, data structures that encode the solutions of the problem. In each generation, a new population is created through the selection and variation of the individuals from previous populations. The individual selection is performed based on the evaluation of the individuals according to a fitness function, and how the individuals are selected and varied depends on the specific algorithm used. Therefore, for using an Evolutionary Algorithm it is necessary to establish the individual encoding, the fitness function and the specific algorithm to be applied.

Regarding the encoding of the individuals, the set of transition rules that make up the CA are directly encoded as vectors consisting of $9M+2$ floating point values in the [0, 1] interval. They are a direct representation of the threshold angles and rules of the automata. The evolutionary algorithm is based on the evaluation, comparison and variation of the individuals of the population. Thus, for the evaluation of each individual it is necessary, firstly, to provide a set of sample images over which the prospective cellular automata can be evaluated, and secondly, to define a fitness function which appropriately reflects how the cellular automata is performing the desired task.

Due to the ever present problem of lack of enough adequately labelled hyperspectral images to be used as training images whenever hyperspectral segmentation processes are addressed, we have decided to explore the possibility of using adequately constructed synthetic RGB images as training samples for the evolutionary process. This solution can prospectively provide several advantages:

1. In the field of hyperspectral imaging, it is difficult to find a set of hyperspectral images labelled with reliable ground truth. Inappropriate labelling will cause errors in the evaluation of the individuals during the evolutionary process. However, using synthetic images, the cellular automata designer completely controls the labelling of the synthetic images, which brings with it a minimization of the evaluation error.

2. Creating synthetic images allows us to train the cellular automata for a given range of desired image features. For example, in the case of segmenting an image by applying a cellular automaton, the level of segmentation can be set depending on the features of the training images and on the labelling of the associated ground truths. Examples of different image features that the cellular automata designer could control are: size of regions, shape of regions, shape of borders among regions, heterogeneity of the pixel spectra belonging to the same region, etc. It means that a cellular automaton can be evolved and set considering a particular final application.

3. As the designed cellular automaton operates with distance measures, which are independent of the dimensionality of the images, it should be possible to train it using images with low dimensionality, for example, RGB.
4. RGB images are much simpler to produce and it takes shorter to process them.
5. Instead of always using the same training images for evaluating all the individuals in the evolutionary process, it is possible to generalize prospective cellular automata through the evaluation of every individual using different training images created at execution time.

There are several different possibilities for measuring the quality of the segmentation obtained after applying the automaton to an image. Here an error measure that considers a balance between the homogeneity of the segmented regions and the discrimination of the different regions within the processed training image was chosen. The fitness function to be minimized is defined as de maximum value between two error measures: the intra-region and the inter-region error:

$$e = \max(\text{intra-error, inter-error}) \tag{11.5}$$

$$\text{intra-error} = \sum_{i=1}^{K} \frac{\max\left(\text{std}\left(p_{i,1}\right),\ldots,\text{std}\left(p_{i,n}\right),\ldots,\text{std}\left(p_{i,N}\right)\right)}{K \cdot 0.5} \tag{11.6}$$

$$\text{inter-error} = 1 - \sum_{i=1}^{K} \left(\sum_{l=1}^{nr_j} \frac{f\left(\alpha_{l,m_j},\alpha_{th}\right)}{nr_i} \right) / K \tag{11.7}$$

being

$$f\left(\alpha_{l,m_j},\alpha_{th}\right) = \begin{cases} 1 & \text{if } \alpha_{l,m_j} > \alpha_{th} \\ 0 & \text{if } \alpha_{l,m_j} \leq \alpha_{th} \end{cases} \tag{11.8}$$

K is the number of different regions in the hyperspectral image, N is the number of bands in the image, $p_{i,n}$ is the reflectance value of the nth band of the pixels belonging to region i, $std(p_{i,n})$ is the standard deviation of $p_{i,n}$, n_{ri} is the number of pixels of region i, $\alpha_{l,m}$ is the spectral angle between pixel l of region i and the average spectrum of region i and α_{th} is a threshold angle value.

Thus, the selected fitness function is an evaluation error e, which is defined as the maximum between two error measures: the intra-region error and the inter-region error (separated by a comma in the formula). The first one provides a measure of the homogeneity of the regions in the image, being a region a set of pixels that should end up with the same label after segmentation. To this end, the maximum standard deviation is computed band by band for all the pixels in the region and the intra-region error is obtained as a sum of these values. Therefore, if all the pixels in a region have the same spectrum, such a region would be completely homogenous and the maximum band by band standard deviation would be zero.

The inter-region error provides a measure of the dissimilarity of two different regions. If all the pixels in the image have the same spectrum, this error is maximized. To compute it, the sum of the result of the following steps for each region is obtained:

1. The average spectrum is calculated.
2. The spectral angle between the average spectrum and the spectrum of the remaining pixels in the image is calculated.
3. The number of angles above a threshold value is computed and normalized with respect to the total number of pixels out of the region under consideration.

Finally, the particular evolutionary algorithm employed in this work is a standard Real Valued Genetic Algorithm (RVGA). Once the evolutionary process finishes, the individual with the highest fitness (minimum error function value) is selected as the resulting CA that will be applied over the image or images to be segmented.

11.3 Validation over synthetic images

To show that the training methodology produces the desired results and in order to be able to quantify how good the results are it is necessary to have a set of hyperspectral test images that are perfectly segmented and labeled. This is not easy and, therefore, as a first approximation, this section presents the results of the operation of ECAS, trained using RGB images, over a synthetic hyperspectral image. In the following two sections, we will consider the application of ECAS to real hyperspectral images.

The quality of the results will be given using a set of very popular metrics that are often used for evaluating classification results. The fact that many authors use them allows comparing these results to others in the literature. These are [16]:

1. *Overall Accuracy (OA)*: it is the percentage of correctly classified pixels

2. *Average Accuracy (AA)*: it is the mean of class-specific accuracies, that is, the mean of the percentage of correctly classified pixels for each class
3. *Kappa coefficient (κ)*: it is the percentage of agreement (correctly classified pixels) corrected by the number of agreements that would be expected purely by chance.
4. *Class-specific accuracies*: they are the percentage of agreement of each class (correctly classified pixels).

As a first example, we consider the performance of the approach over perfectly labeled and controlled synthetic images. They were constructed using five 64-band base spectra that were spatially corrupted by noise and artifacts. Figure 11.3 displays the result of applying the CA resulting from the evolutionary process presented in the previous sections over a set of simple RGB images and the segmentation and classification algorithms over a synthetic hyperspectral image containing 5 different classes. The top image of the figure corresponds to a 2D transformation of the synthetic 64-band hyperspectral where each pixel corresponds to the angle between the spectrum of the pixel and a reference spectrum with all of its bands at the maximum value. Here it is easy to see that the image is not homogeneous at all, it is actually quite a complicated image. The center image displays the ground truth. Finally, the bottom image shows the final classification obtained with the approach presented here.

A simple visual analysis indicates that the final classification image and the ground truth are very similar, with the same areas clearly defined. After iterating the CA 100 times over the synthetic image, the following performance results were obtained:

Table 11.1. Performance results for synthetic image

OA	98.8 %
AA	96.4%
Kappa	98.2%
CLASS SPECIFIC ACCURACIES	
Class 1	99.8%
Class 2	97.4%
Class 3	98.0%
Class 4	98.4%
Class 5	88.5%

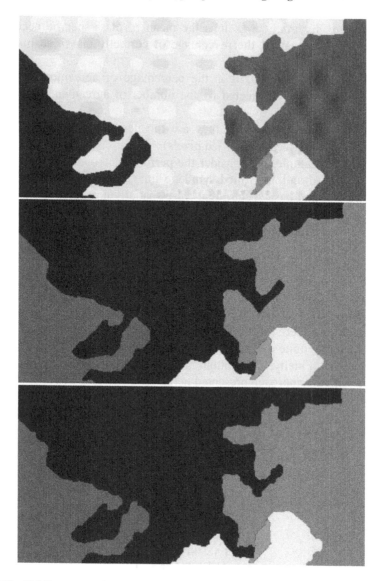

Fig. 11.3 From top to bottom: 2D angle wise transformation of a synthetic 64-band
hyperspectral image, ground truth and final classification

These data imply that the classification accuracy is very high using an
image whose classification results are perfectly known. In fact, most of the
small classification error present stems from pixels right on the borders,
which is something that will have to be addressed in further versions of the

segmentation algorithm. Notwithstanding this, the most relevant result here is that from a spectrally inhomogeneous hyperspectral image, as shown in the top image of Figure 11.3, the CA that was evolved using RGB training samples is able to provide a very homogeneous pre-segmented image that is really easy to segment and classify.

11.4 Application example: Salinas

In this section, we present the results provided by the ECAS algorithm described in the previous sections for segmenting in an unsupervised manner a well-known hyperspectral image: the AVIRIS Salinas image.

To train ECAS we generated at runtime synthetic images similar to those shown in Figure 11.4. Every candidate automaton in the population was evaluated using a different image, although all of them had similar types of features corrupted by noise and artifacts. We have used a CA as the one described in section 11.2.1 with a total of 20 rules. As each rule has 9 components and we have two additional parameters corresponding to the threshold angles, we end up with 182 real valued parameters for each automaton.

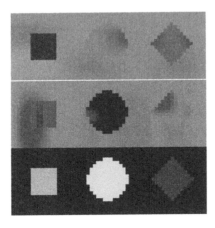

Fig. 11.4. Top: Training RGB image. Bottom: Ground truth

Figure 11.5 displays the evolution of the fitness of the best individual and the average fitness of the population for one of the evolutionary processes performed in this case. The fluctuations that can be observed in the fitness curves are due to the fact that the evaluation of the individuals is carried out over different images. However, the global trend followed by the curves is clear, and the error decreases as expected.

The results of the segmentation obtained with the best CA after evolution are displayed in Table 11.2 and in the rightmost part of Figure 11.6. They are visually quite acceptable and show that the selected CA has obtained a set of rules that allow it to clearly segment the image into the relevant areas except for a slight over-segmentation in areas 9 and 2, for each one of which the CA delimits two categories, and some noisy pixels in the borders of some areas. It is difficult to really know if the aforementioned over-segmentation represents some relevant feature of the image, as the only information we have on the area is the ground truth image displayed in the left image of Figure 11.6. However it must be pointed out that such areas do present different spectra in the original image and, consequently, from a strict segmentation point of view, they are correctly segmented. On the other hand, the noisy pixels in the borders of some areas have to do with imprecise delimitation of the areas in the ground truth (borders of fields that do not contain plants and are taken as paths). However, to really be able to quantify the goodness of the segmentation we have resorted to the Overall Accuracy index (the percentage of correctly classified pixels), in this case for each one of the areas with the same label in the ground truth image. These OA values are presented in Table 11.1 and they lead to a global OA of 94.1%.

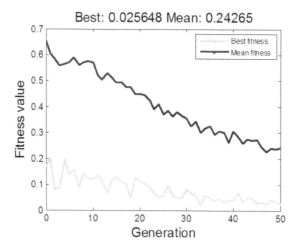

Fig. 11.5. Fitness function value (minimized) during one of the evolutionary processes

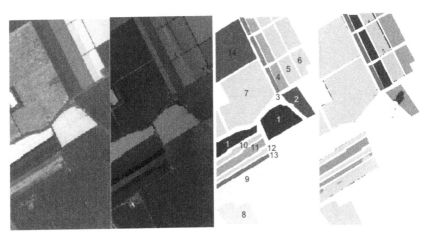

Fig. 11.6. From left to right: 2D angular transformation of the AVIRIS scene, 2D angular transformation of the hyperspectral cube resulting from iterating the CA over the image 100 times; ground truth (the content of each area is indicated in Table 11.2) and the result of the unsupervised CA based segmentation process (the colors do not correspond to labels, they just indicate homogeneous regions or segments as determined by the CA)

It is important to note that the unsupervised segmentation procedure presented here seeks to provide a single label in terms of a single spectrum to represent each area; however, we are talking about segmentation, not classification. As a consequence the representative spectrum may not be the one corresponding to the ground truth label and when two separate areas in an image that contain the same type of cover are segmented they may be assigned different labels (much in the same way as in the case of watershed based segmentation algorithms). That is, for instance, in an area where there are very small plants each surrounded by a lot of soil, the representative spectrum obtained in an unsupervised manner may be very close to soil and if one takes it at face value, without a posterior labeling or classification process, this may lead to confusion between areas that present this characteristic. Again, we are segmenting and if a posterior classification process is desired, this can be achieved selecting over the original hyperspectral cube a representative spectrum for each of the areas the segmentation algorithm has defined (maybe the average spectrum of the area) and use another process to decide what that spectrum corresponds to or whether it is a composition of several end-members.

Table 11.2. Overall accuracy (%)

Area	Cover	OA (%)
1	Brocoli	99,7
2	Fallow	82,4
3	Fallow rough plow	100
4	Fallow smooth	98,4
5	Stubble	93,8
6	Celery	99,5
7	Grapes	98,7
8	Soil vineyard develop	100
9	Corn senesced, green weeds	70,3
10	Lettuce romaine 4wk	94,8
11	Lettuce romaine 5wk	100
12	Lettuce romaine 6wk	97,8
13	Lettuce romaine 7wk	83,4
14	Vineyard	99,3

11.5 Application example: Pavia

Another very popular benchmark in hyperspectral image segmentation is that of the University of Pavia campus. This is the second application example considered in order to show the performance of the ECAS algorithm trained over lower dimensional images when applied to real hyperspectral cases.

The University of Pavia image is a 610x340 pixel image that contains 103 spectral bands in the 430 to 860 nm interval where 9 different classes haven hand labeled (they are listed in Table 11.3).

Again, as in the previous example, a group of synthetic images similar to those shown in Figure 11.7 were generated at runtime to train ECAS. As in the previous case, every candidate automaton in the population was evaluated using a different image, although all of them had similar types of features corrupted by noise and artifacts.

Figure 11.8 shows the results of applying the resulting CA to the segmentation of the University of Pavia image.

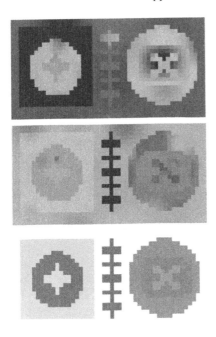

Fig. 11.7. Top: Training RGB images. Bottom: Ground truth

Image (a) displays a 2D angular transformation of the original hyperspectral image where each pixel corresponds to the angle between the spectrum of the pixel and a reference spectrum with all of its bands at the maximum value. Image (b) corresponds to the ground truth where the 9 classes that are considered have been marked in different colors. This image is only partially labeled and, even though many authors use it in order to perform comparisons; the labeling is at least dubious in some cases, especially in the areas corresponding to shadows. Image (c) corresponds to the final image where only the labeled areas are displayed while image (d) contains the classification of all the areas in the original image. Again, a simple visual comparison of the ground truth images and the bottom left image shows the high degree of success that has been achieved.

The goodness of the classification results is quantitatively confirmed by the class-specific accuracy presented in Table 11.3. The main differences between the images are in very specific areas. For instance, it must be pointed out that the lower accuracy obtained for class 9 (shadows) is mainly because it is quite a confusing label. In fact, this label does not really correspond to a specific spectrum but to several different spectra that correspond to different types of ground covers but with lower intensities (the amount of sunlight impacting these areas is less and thus the reflection is also less intense).

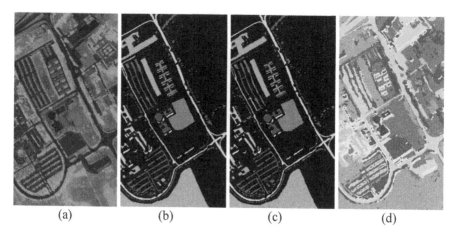

(a) (b) (c) (d)

Fig. 11.8. (a) 2D angular transformation of the PAVIA image. (b) Ground truth. (c) Final image provided by the two-step algorithm considered here showing only the labeled areas. (d) Final classification results provided by the algorithm for the whole image

Table 11.3. Class-specific accuracy (%) for the University of Pavia image

Area	Class	OA(%)
1	Asphalt	94,63
2	Meadows	98,92
3	Gravel	98,57
4	Tree	86,29
5	Metal sheets	95,17
6	Bare soil	99,66
7	Bitumen	96,62
8	Bricks	91,01
9	Shadows	84,16

In order to compare the overall classification results obtained by ECAS over this image to those of other authors, three reference algorithms were selected. To provide reference values, a purely spectral algorithm based on a pixel-wise SVM has been used (the data comes from [16]). In addition, two spatial-spectral algorithms were considered. The first one is the EMP (extended morphological profiles), originally presented by Plaza et al. in

[13] and the second one is a watershed transformation-based algorithm presented by Tarabalka et al. in [16], which has been labeled as W-RCMG in Table 11.4. This algorithm considers a SVM based classification with majority vote where the watershed pixels are assigned to the regions with the closest median before the majority vote is performed (using a WHEDS approach) and which uses a Robust Color Morphological Gradient (RCMG) to obtain the gradient image. It is this second algorithm the one that provided the best results in the comparison the authors performed over the same image in [16]. As it can be observed in the table, the ECAS+SVM combination outperforms the other three algorithms in all the measurements for this particular benchmark image, showing the validity of the proposed approach.

Table 11.4. Overall accuracy (%) for different algorithms

	ECAS+SVM	Pixel-wise SVM	EMP	W-RCMG
Kappa	**94,99**	75,86	80,86	81,30
OA	**96,22**	81,01	85,22	85,42
AA	**93,89**	88,25	90,76	91,31

11.6 Conclusions

This chapter has presented an evolutionary approach to obtain the rule sets for Cellular Automata (CA) based structures that must carry out spatial-spectral operations over hyperspectral images. Its main objective was to present a methodology to address the ever present problem within the field of the lack of appropriately and reliably labeled images that can be used for training and adequately evaluating CAs. That is, it is hard to construct large enough sets of images that can be used to evaluate the fitness of the candidate CAs during evolution. Additionally, manipulating these high dimensional images implies a lot of computation that slows down the training process. The problem has been addressed through the adoption of a dimension independent distance metric within the operation of the CAs. This allows evolving the automata using perfectly labeled synthetic images of any dimensionality. Therefore, for the sake of simplicity one can choose images of a much lower dimensionality for the training process than those over which the resulting CA will be applied.

Another aspect this approach is careful about is that of the preservation of the large amount of information real spectral provide when carrying out segmentation processes. In this line, the CA based method presented in this chapter works with and modifies the spectral information, without projecting it onto lower dimensionalities at any point. This leads to the production of complete hyperspectral images as a result of the segmentation process where the different regions have been spectrally homogenized. After this process they can be easily labeled through any algorithm for spectral classification.

The results of a series of tests carried out over synthetic and real benchmark images have been carried out and their results show that this approach is very competitive when compared to other algorithms presented in the literature.

References

[1]   C. Chang and C. Lin, "LIBSVM – a library for support vector machines". *ACM Transactions on Intelligent Systems and Technology (TIST)*, Vol. 3, No. 2, pp. 27, 2011.

[2]   Z. Chuanwu, "Performance analysis of the CPLD/FPGA implementation of cellular Automata". *IEEE Trans. of the 2008 International Conference on Embedded Software and Systems Symposia (ICESS'08)*, pp. 308–311, 2008.

[3]   A. Darwish, K. Leukert, and W. Reinhardt, "Image segmentation for the purpose of object-based classification". *IEEE International Geoscience and Remote Sensing Symposium*, Vol. 3, No. C, pp. 2039–2041, 2003.

[4]   R. Duro, F. Lopez-Pena, and J. Crespo, "Using Gaussian synapse ANNs for hyperspectral image segmentation and endmember extraction". *Computational Intelligence for Remote Sensing*, Springer, pp. 341–362, 2008.

[5]   O. Eches, N. Dobigeon, and J. Y. Tourneret, "Markov random fields for joint unmixing and segmentation of hyperspectral images". *2nd Workshop on Hyperspectral Image and Signal Processing: Evolution in Remote Sensing (WHISPERS'10)*, pp. 1–4, 2010.

[6]   A. Farag, R. M. Mohamed, and A. El-Baz, "A unified framework for MAP estimation in remote sensing image segmentation". *IEEE Transactions on Geoscience and Remote Sensing*, Vol. 43, no. 7, pp. 1617–1634, 2005.

[7]   G. Flouzat, O. Amram, and S. Cherchali, "Spatial and spectral segmentation of satellite remote sensing imagery using processing graphs by mathematical morphology". *IEEE International Geoscience and Remote Sensing Symposium (IGARSS'98)*, Vol. 4, pp. 1769–1771, 1998.

[8]   N. Ganguly, B. K. Sikdar, A. Deutsch, G. Canright, and P. P. Chaudhuri, "A survey on cellular automata". *Technical report, Centre for High Performance Computing, Dresden University of Technology*, pp. 1–30, 2003.

[9]   D. B. Heras, F. Arguello, J. L. Gomez, J. A. Becerra, and R. J. Duro, "Towards real-time hyperspectral image processing, a GP-GPU implementation of target identification". *IEEE 6th International Conference on Intelligent Data Acquisition and Advanced Computing Systems (IDAACS'11)*, Vol. 1, pp. 316–321, 2011.

[10] J. Li, J. M. Bioucas-Dias, and A. Plaza, "Hyperspectral image segmentation using a new Bayesian approach with active learning. *IEEE Transactions on Geoscience and Remote Sensing*, Vol. 49, no. 10, pp. 3947–3960, October 2011.

[11] P. Li and X Xiao, "Evaluation of multiscale morphological segmentation of multispectral imagery for land cover classification". *IEEE International Geoscience and Remote Sensing Symposium (IGARSS'04)*, Vol. 4, No. C, pp. 0–3, 2004.

[12] M. Pesaresi and J. A. Benediktsson, "A new approach for the morphological segmentation of high-resolution satellite imagery". *IEEE Transactions on Geoscience and Remote Sensing*, Vol. 39, No. 2, pp. 309–320, 2001.

[13] A. Plaza, J. A. Benediktsson, J. W. Boardman, J. Brazile, L. Bruzzone, G. Camps-Valls, J. Chanussot, M. Fauvel, P. Gamba, A. Gualtieri, M. Marconcini, J. C. Tilton, and G. Trianni, "Recent advances in techniques for hyperspectral image processing". *Remote Sensing of Environment*, Vol. 113, pp. S110–S122, July 2009.

[14] B. Priego, D. Souto, F. Bellas, R. J. Duro, "Hyperspectral image segmentation through evolved cellular automata". *Pattern Recognition Letters*, Vol. 34, No. 14, pp. 1648–1658, October 2013.

[15] P. Quesada-Barriuso, F. Argüello, and D. B. Heras, "Efficient segmentation of hyperspectral images on commodity GPUs". *Advances in Knowledge Based and Intelligent Information and Engineering Systems*, Vol. 243, pp. 2130–2139, 2012.

[16] Y. Tarabalka, J. Chanussot, and J. A. Benediktsson, "Segmentation and classification of hyperspectral images using watershed transformation". *Pattern Recognition*, Vol. 43, No. 7, pp. 2367–2379, 2010.

[17] J. C. Tilton, "Analysis of hierarchically related image segmentations". *IEEE Workshop on Advances in Techniques for Analysis of Remotely Sensed Data*, Vol. 00, No. C, pp. 60–69, 2003.

[18] T. Veracini, S. Matteoli, M. Diani, and G. Corsini, "Robust hyperspectral image segmentation based on a non-Gaussian model". *2nd International Workshop on Cognitive Information Processing (CIP)*, pp. 192–197, 2010.

Biography

Blanca María Priego Torres received the title of Telecommunications Engineer in 2009 from the University of Granada, Spain. In 2011, she obtained the Master's Degree in Information and Communications Technologies in Mobile Networks from the University of A Coruña, Spain. She is currently working in her PhD as a member of the Integrated Group for Engineering Research at the University of A Coruña. Her research interests include new neural-network structures and hyperspectral image analysis applied to the Industrial and Naval Field.

Daniel Souto received the title of Industrial Engineer in 2007 from the University of A Coruña, Spain. He is currently working towards a Ph. D. degree in the Department of Industrial Engineering at the same University. He is currently a researcher at the Integrated Group for Engineering

Research. His research activities are related to automatic design and mechanical design of robots and intelligent measurement systems.

Francisco Bellas is a *Profesor Titular* at the University of A Coruña, Spain. He received the B.S. and M.S. degree in Physics from the University of Santiago de Compostela, Spain, in 2001, and a PhD in Computer Science from the University of A Coruña in 2003. He is a member of the Integrated Group for Engineering Research at the University of A Coruña. His main research activities focus on evolutionary robotics, collective intelligence and neuroevolutionary algorithms.

Richard J. Duro received a M.S. degree in Physics from the University of Santiago de Compostela, Spain, in 1989, and a PhD in Physics from the same University in 1992. He is currently a Full Professor in the Department of Computer Science and head of the Integrated Group for Engineering Research at the University of A Coruña. His research interests include higher order neural network structures, multidimensional signal processing and autonomous and evolutionary robotics.

Index

About the editors

Vladimir Haasz finished Czech Technical University (CTU) in Prague, Faculty of Electrical Engineering (FEE) in 1972 (branch "Technological Cybernetics"), and since that year he has been with Department of Measurement. In 1977 he obtained his Ph.D. degree and in 1991 he defended his habilitation thesis. In 1994–1995 he spent the half-year at ETH Zürich as a senior researcher. He was named as Full Professor of Measurement Technology in 1999. He managed the Department of Measurement at CTU-FEE 1997–2008 and 2011–2014.

Vladimir Haasz is a member of TC 12 "Quantities and Values" of the Czech Institute of Standardization, of IMEKO (International Measurement Confederation) TC-4 – Measurement of Electrical Quantities, and of the editorial board of International Journal of Computing. He has been the honorary member of the Scientific Counsel of CTU 2011–2013.

He is interested in the field of measuring systems of electrical quantities, sampling methods of measurement of non-harmonic waveforms, and in the last years especially in testing of dynamic quality of AD modules and their EMC. He gives lectures on the basic courses "Electrical Measurement and Instrumentation" and "Sensors and Measurement", and on the optional course "Advanced Instrumentation".

Kurosh Madani graduated in fundamental physics in June 1985 from PARIS 7 – Jussieu University (Paris, France), he received his MSc. in Microelectronics and complex processors' architecture from University PARIS 11 (PARIS-SUD), Orsay, France, in September 1986.

He received his Ph.D. in Electrical Engineering and Computer Sciences from University PARIS 11 (PARIS-SUD), Orsay, France, in February 1990. From 1989 to 1990, he worked as assistant professor at Institut d'Electronique Fondamentale (Institute of Fundamental Electronics) of PARIS 11 University and CNRS (National Center of Scientific Research), Orsay, France.

In 1990, he joined Creteil-Senart Institute of Technology of University PARIS-EST Créteil (UPEC), Lieusaint, France, where he worked from 1990 to 1998 as assistant professor. In 1995, he received the DHDR Doctor Hab. degree (senior research doctorate degree) from UPEC.

Since 1998 he has been working as Chair Professor in Electrical Engineering of Senart Institute of Technology of UPEC.

From 1992 to 2000 he was creator and head of DRN (Neural Networks Division) research group in LERISS laboratory of UPEC. From 2001 to 2004 he has been head of Intelligence in Instrumentation and Systems Laboratory of UPEC, located at Senart Institute of Technology. Director of SCTIC research division, one of the two research divisions of Images, Signals and Intelligent Systems Laboratory (LISSI / EA 3956) of UPEC from 2005 to 2009, he is Vice-director of LISSI since 2009.

Concerning his research interests, he has worked on both digital and analog implementation of massively parallel processors arrays for image processing by stochastic relaxation, electro-optical random number generation, and both analog and digital Artificial Neural Networks (ANN) implementation.

Author and coauthor of more than 300 publications in international scientific journals, books (Springer, Kluwer, etc.), international conferences' and symposiums' proceedings, he has been regularly invited as key-note and invited lecture by international conferences and symposiums (IEEE, IFAC, etc.). His current research interests include

- Bio-inspired Artificial systems' modeling and implementation,
- self-organizing, modular and hybrid neural based information processing systems and their software and hardware implementations,
- Soft-Computing based complex applications: Automated negotiation mechanisms and systems, neural based fault detection and diagnosis systems, design and implementation of real-time neuro-control, etc.
- humanoid robotics
- collective robotics and collective intelligence
- Intelligent machines & systems

Since 1996 he is life-member (elected permanent Academician) of International Informatization Academy. Since 1997, he is also elected permanent Academician (life-member) of International Academy of Technological Cybernetics.

Lightning Source UK Ltd.
Milton Keynes UK
UKOW06n0609081215

264273UK00001B/47/P